A Guide to Feedba

Do you need to understand feedback? Perhaps you're a little rusty on theory basics? Dig in to this self-contained guide for an accessible and concise explanation of the fundamentals.

- Distills the relevant essence of linear system theory, calculus, differential equations, linear algebra, basic physics, numerical methods, and complex analysis and links them back to an explanation of feedback theory.
- Provides a tight synthesis of analytical and conceptual understanding.
- Maintains a focus on common use cases.

Whether you are a struggling undergraduate, a doctoral student preparing for your qualifying exams, or an industry practitioner, this easy-to-understand book invites you to relax, enjoy the material, and follow your curiosity.

JOEL L. DAWSON is an entrepreneur and former MIT professor. He is a 2009 recipient of the PECASE Award, the highest honor bestowed by the US government on young scientists and engineers. His last start-up company, Eta Devices, Inc., was a Technology Pioneer of the 2015 World Economic Forum and acquired by Nokia in 2016. His current start-up company is TalkingHeads Wireless, Inc.

"Feedback theory is an intrinsically mathematical discipline in which one can feel either submerged by formulae or driven to use blind computer simulations that hide insight. Dawson's approach is to extract visceral meaning out of this tangle, arguing that a deep understanding of dynamic stability criteria can free the designer from "equational overload" and lead to incisive selection of the right mathematical tool for the job at hand."

Stephen D. Senturia, *Massachusetts Institute of Technology*

"Feedback is perhaps the most foundational concept for electronics and control systems in general, but it is often covered for specific circuits for the former, and in terms of theoretical concepts for the latter. This book provides us with a unique perspective on how feedback theory in general relates to practical systems and electronics applications."

Larry Pileggi, *Carnegie Mellon University*

A Guide to Feedback Theory

Joel L. Dawson
TalkingHeads Wireless, Inc.

CAMBRIDGE
UNIVERSITY PRESS

CAMBRIDGE
UNIVERSITY PRESS

University Printing House, Cambridge CB2 8BS, United Kingdom

One Liberty Plaza, 20th Floor, New York, NY 10006, USA

477 Williamstown Road, Port Melbourne, VIC 3207, Australia

314–321, 3rd Floor, Plot 3, Splendor Forum, Jasola District Centre, New Delhi – 110025, India

79 Anson Road, #06–04/06, Singapore 079906

Cambridge University Press is part of the University of Cambridge.

It furthers the University's mission by disseminating knowledge in the pursuit of education, learning, and research at the highest international levels of excellence.

www.cambridge.org
Information on this title: www.cambridge.org/9780521199216
DOI: 10.1017/9781139018289

© Cambridge University Press 2021

First published 2021

Printed in the United Kingdom by TJ Books Limited, Padstow Cornwall

A catalogue record for this publication is available from the British Library.

ISBN 978-0-521-19921-6 Hardback
ISBN 978-0-521-15393-5 Paperback

For Elías

Contents

Preface

This is a book about classical feedback control and, along the way, a review of the foundational linear system theory that is at the core of the electrical engineering discipline. But there are many, many books on these subjects. What makes this book different?

The answer is that this book, in its structure, content, and style, targets four very specific groups of people. In no particular order:

- **The drowning undergraduate.** In this, I take inspiration from *Signals and Systems Made Ridiculously Simple* by Zoher Z. Karu (ZiZi Press, 1995). While very different in style, his book was a godsend to us undergraduates at MIT for its simple, short, clear explanations. More deeply, the book unobtrusively guided students *toward* what is important and *away* from what is not. With years to reflect, I look back on that unpretentious volume with affection, gratitude, and respect.

- **The superstar undergraduate.** Perhaps you are absolutely crushing your exams and reaping praise from your professors accordingly. This is a good feeling, and sometimes a just reward for hard work. But a peculiarity of our species can manifest quite strongly here: dominating exam performance is some mix of (1) deep understanding of the material, on one hand, and (2) on the other hand, a virtuosity in pure test taking that owes much to a kind of supercharged social intelligence. What is really astonishing, especially in a competitive, "elite" college environment, is just how far the latter can take you with little development of the former. Given how we structure higher education, an overreliance on (2) in some students is inevitable and natural and to be neither condemned nor praised. But if you have taken up this book, take the opportunity in reading it to do a little self-assessment and reflection. It is the deep understanding of (1) that you need to build useful and interesting machines.

- **The graduate student preparing for their qualifying exams.** Many doctoral programs kick things off in the first or second year with a weighty set of exams designed to probe a candidate's understanding of basic undergraduate material. The weeks or months preparing for a doctoral exam can be extremely rewarding: one discovers that the "basic" undergraduate material is both broad and deep. Moreover, this is a time when many emerging scholars and practitioners come into their own as true self-learners. When I had the first discussions with Julie Lancashire at Cambridge University Press about this book, my original inspiration was to write the book I wished had existed when I was studying for my own qualifying exams. I have tried to stay true to that spirit throughout.
- **A subset of industry practitioners.** Technology in mature and high-volume industries doesn't change very often. In cellular wireless networks, for example, at the time of this writing, the dominant power amplifier architecture continued to be the Doherty structure, invented at Bell Labs in 1936. It is very possible to be a professional in mature fields *and build a successful career*, while having only a loose grasp of the underlying physics. Trade-offs in the performance space can be memorized ("You know, Bill, if you make it faster, it's definitely going to be less stable. There are always trade-offs, my friend, always trade-offs"), and vocabulary can be used accurately by people who have a superficial understanding of where it all comes from. The danger here is that technological paradigm shifts do happen from time to time. If you understand the foundations of your field, you will be able to acquire a similar understanding of the new paradigm through diligent self-study. If your understanding has always been superficial, however, you are helpless and exposed in the face of major change. This can be quite scary.[1] If you find yourself vulnerable to this type of situation, the goal of this book is to be a friendly example of a new type of understanding. Relax and have fun with the material, and allow yourself to follow your curiosity.

My hope for the reader is that you find this book as enjoyable and informative to read as it was to write.

---

[1] If you are very lucky, old enough, and happen to look the part, you might be able to pull a convincing turn as a curmudgeon of the Old School who simply prefers things that have Stood the Test of Time.

Acknowledgments

First I would like to acknowledge Julie Lancashire, Sarah Strange, and Julia Ford, members of the editorial staff at Cambridge University Press. The collaborative, creative energy of my conversations with Julie was fuel over the more than a decade that this project has required. Both Julie and Sarah kept in touch with me over a years-long stretch when I was building a start-up company, and it must have seemed like I would never deliver a polished manuscript. Julia has been instrumental in getting the book across the finish line, and the figures in this text have benefited tremendously from her sharp eye for detail. I am also indebted to multiple anonymous reviewers. They clearly understood what I was trying to accomplish with this unusual book and gave tremendous suggestions for improvement to that end.

I am immensely grateful to Prof. James K. Roberge (1938–2014). At various times he was a mentor, teacher, friend, and finally colleague to me on the faculty at the Massachusetts Institute of Technology. I first met him at a time in my life when my curiosity about feedback and electrical engineering far exceeded my knowledge. It may be that there was no better person in the world to meet in that circumstance than him.

Finally, I give heartfelt thanks to my family and friends. They have been a source of encouragement, inspiration, and support through many ups and downs over the years. This book would not have been possible without them.

1

Linear Systems: What You Missed the First Time

For an engineer, math is a language of unusual expressive power and concision. The first time that you studied differential equations, however, chances are high that this escaped you. That is natural. The purpose of this chapter is to help you to get in touch with the meaning behind the math of linear, time-invariant (LTI) systems.

1.1 Differential Equations Are a Natural Way to Express Time Evolution

Feedback systems are dynamic: ultimately, we are interested in the evolution of their state over time. Frequency-domain tools like the Laplace transform are wonderful aids for analysis, but before revisiting those let's examine the basic differential equation. We will see that despite appearances, it is quite a natural way to describe the time evolution of a dynamic system.

1.1.1 A First-Order System

Consider the RC circuit shown in Figure 1.1. The situation is that the capacitor has a charge Q_0 while the switch is open. The switch is then closed, which connects a resistor of value R across the terminals of the capacitor. To determine what happens next, we have at least two approaches. First, we can argue on physical grounds that eventually the capacitor must completely discharge, leaving the zero voltage across the capacitor. A sophisticated observer might even point out that the capacitor will never *fully* discharge, or alternatively, that a complete discharge would take an infinite amount of time. A second approach is to not bother at all with "intuitive" reasoning and

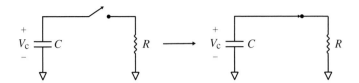

Figure 1.1 Discharging a capacitor through a resistor. The capacitor has an initial charge $Q_0 = CV_0$ and begins to discharge when the switch is closed.

physical insight. We just write down the differential equation governing the system and derive an expression for $Q(t)$ in exquisite detail.

The problem with the first approach alone is that it does not always yield the level of detail that we might require. While we can say that it will take longer to discharge if the resistor R is bigger, we are helpless to say exactly how long it will take the capacitor to lose 90 percent of its charge, for example. The problem with the second approach alone is that without physical insight, the student can never progress beyond solving little, well-packaged problems with neat answers. *If the engineer is ever to unleash their creativity to invent, design, build, and discover new things, they will be powerfully aided by understanding the yin-and-yang interplay between physical insight and mathematical analysis.*

The simple discharging of a capacitor is a great way to start understanding this balance. Starting with the initial conditions, we have an open switch and a capacitor with charge Q_0. The physical meaning of capacitance is it tells us how much charge we must supply if we are to establish a potential difference between two conductors. The greater the capacitance, the more charge we must supply to establish a given potential difference. This is beautifully and succinctly captured by the constitutive law for capacitors, $Q = CV$. We know therefore that before we throw the switch, the voltage across the capacitor terminals is $V_0 = Q_0/C$.

When we do throw the switch, we have a new constitutive relation to satisfy, namely, Ohm's law. In the first instant after the switch is closed, the charge flows through the resistor at a rate of $I = V_0/R = Q_0/RC$ Coulombs per second. But as soon as the first tiny bit of charge is removed from the capacitor, the voltage across the capacitor goes down, causing the current to decrease, which nevertheless continues to remove charge from the capacitor, and so on and so forth. At this point we have a good physical understanding of what is happening. How can we ever find out *exactly* how the charge decays with time? The approach is to express our physical insight mathematically. Almost immediately, though, we run head-first into the problem of how to deal with the progression of time. Things would be easier if time moved forward in discrete chunks. We cannot help, for example, talking about the current flow "in the first instant" after the switch is closed. Unfortunately, we know (or, at least, have no reason to doubt) that time moves forward in a continuous progression. Before we've written our first equation, then, we have a seemingly good reason to despair.

The key insight is to realize that for any "continuous" variable, there is a level of granularity beyond which a discretized representation is, for all *practical* purposes, indistinguishable from a continuous one. We know that water, for example, is composed of discrete water molecules, yet to our unaided senses the granularity is so fine as to be indistinguishable from a continuous liquid. The time variable is no different. Suppose that nature actually moved in steps of one femtosecond (10^{-15} seconds). Would we be any the wiser, even if using the fastest oscilloscopes available at the time of this writing?[1] There might be other ways of telling if nature is secretly discretizing time, but to an engineer with an oscilloscope, there is no practical difference between a universe that discretizes time in one-femtosecond chunks and one that moves forward continuously in time. This critical realization helps us to move forward, and ultimately leads us to the shorthand that we now know as differential equations.

Returning to our problem, we might consider breaking time up into tiny chunks of duration Δt. If we know the charge on the capacitor at time t, we ask "What is the charge at time $t + \Delta t$?" If we know the answer to this question in general, and we know the answer at time $t = 0$ (or some other initial time), then we know the answer for all time. So we write

$$Q_c(t + \Delta t) = Q_c(t) - I(t)\Delta t. \qquad (1.1)$$

That is, the charge at the next instant is equal to the charge at the current instant, minus the charge that was bled off in one interval of time due to the current at time t. What is value of Δt? At this point we don't bother about it. We keep firmly in our mind that it is small enough so as to be indistinguishable from continuous time, and don't go back and pick a value of Δt until we need numerical answers. And what about the fact that in the truly continuous system, I does not stay constant over any interval of time? It is true that this will introduce an error. What is important is that we can make this error arbitrarily small by making Δt as small as we like. Remember, the goal is *not* to come up

---

[1] The Agilent DSO91304A Infiniium oscilloscope samples at a "pedestrian" 40 GSamples/ second, or once every 25,000 femtoseconds.

with a model that is as accurate as nature is. That is impossible. We need only be as accurate as we can conceivably measure.

Now, since voltage is what we actually measure, we recast Eq. 1.1 in terms of the capacitor voltage

$$CV_c(t + \Delta t) = CV_c(t) - I(t)\Delta t. \tag{1.2}$$

We seize on the fact that $I(t)$ is linked to V_c through Ohm's law: $I(t) = V_c/R$. Substituting and gathering terms, we arrive at

$$\frac{V_c(t + \Delta t) - V_c(t)}{\Delta t} + \frac{1}{RC}V_c(t) = 0. \tag{1.3}$$

And now we appear to be stuck. There is no way to derive an expression for $V_c(t)$ that satisfies this equation. The best we can do is guess at a solution, plug it in, and check to see if it "works" by resulting in an equation that is self-consistent. What is a good guess?

The good news is that we needn't guess blindly. On physical grounds, we expect $V_c(t)$ to decay with time; we expect that the rate of decay will slow with time; we expect it to asymptotically approach zero. An inspired guess, drawn from an admittedly large number of possibilities, is $V_c(t) = V_0 a^{n \cdot \Delta t}$. In this solution, n is an integer index that steps us forward in time, Δt is our time increment, and a is a key parameter. If $|a^{\Delta t}| < 1$, we satisfy all of the conditions we set forth. If we have chosen correctly, the equation will determine the value of $a^{\Delta t}$ unambiguously. This in turn validates our initial guess.

Plugging into Eq. 1.3, we have

$$\frac{V_0 a^{(n+1) \cdot \Delta t} - V_0 a^{n \cdot \Delta t}}{\Delta t} + \frac{1}{RC}V_0 a^{n \cdot \Delta t} = 0. \tag{1.4}$$

A factor of $V_0 a^{n \cdot \Delta t}$ appears in all terms. Dividing both sides of the equation by $V_0 a^{n \cdot \Delta t}$ and simplifying leads to

$$a^{\Delta t} = 1 - \frac{\Delta t}{RC}. \tag{1.5}$$

This is a critical juncture in our development. In some ways, once we have $a^{\Delta t}$ we are done. Equation 1.4 is a first-order polynomial in $a^{\Delta t}$, and in Eq. 1.5 we have an equation that gives us the roots of that polynomial. We'll see polynomials like this again when we look at discrete-time systems starting in Section 1.7. For now, the *only* reason we're continuing from here is that this is not a discrete-time system, and so we must examine the implications of allowing Δt to become arbitrarily small. Continuing, Eq. 1.5 allows us to write

$$\ln a = \frac{1}{\Delta t} \ln\left(1 - \frac{\Delta t}{RC}\right). \tag{1.6}$$

Now, what do we mean when we insist that Δt is small? It is actually meaningless to insist that Δt be "small" in an abstract sense. We must instead specify its smallness in comparison to something. In this problem, suppose that we say that Δt is small compared to the quantity RC. Why does this make sense? Rewriting Eq. 1.3 slightly, we have

$$\frac{V_c(t + \Delta t) - V_c(t)}{V_c(t)} = -\frac{\Delta t}{RC}. \tag{1.7}$$

Put into words, Eq. 1.7 says that $\Delta t \ll RC$ is equivalent to saying that the fractional change in V_c during any given time step is small. This is exactly what we should hope for if we expect to better approximate a continuously evolving system by shrinking the increment Δt.

Now we employ a trick that is very common in all the disciplines of engineering and science. On conditions such as $\Delta t \ll RC$, it is natural to substitute for $f(x_0 + \Delta x)$ a *polynomial expansion*:

$$f(x_0 + \Delta x) \approx a_0 + a_1 \cdot (\Delta x) + a_2 \cdot (\Delta x)^2 + a_3 \cdot (\Delta x)^3 + \cdots \tag{1.8}$$

We like polynomial expansions because they are easy to realize in computational hardware: if you can multiply and add, you can work with polynomials. This is depressingly untrue of transcendental functions like the logarithms, exponentials, and trigonometric functions that surface with such persistence in the analysis of linear systems. When you throw in the condition that $x_0 \ll \Delta x$, the good news just gets better in that you can get excellent numerical accuracy despite truncating the polynomial expansion to a finite number of terms. In fact, we often take $x_0 \ll \Delta x$ to mean that the original function can be well approximated with only *two* terms:

$$f(x_0 + \Delta x) \approx a_0 + a_1 \cdot (\Delta x). \tag{1.9}$$

This happy circumstance is extremely convenient for hand analysis. You may remember this trick as "linearization."

But let's not jump ahead. Let's conservatively "guess" that for our purposes the logarithm can be adequately captured by a third-order polynomial expansion. We'll then check later to see if it introduces unacceptable numerical error. There are many techniques for fitting polynomials. For the function $\ln(1+\Delta x)$, the author chose for data points $\Delta x \in [10^{-4}, 10^{-3.5}, 10^{-3}, 10^{-2.5}, 10^{-2}]$, which are all conspicuously small compared to 1. An elementary least-squares fit[2] results in the polynomial substitution for $\ln(1 + \Delta x)$

---

[2] For the interested reader, an excellent treatment of least-squares fits can be found in Gilbert Strang's *Introduction to Linear Algebra*, 4th ed. (Wellesley, MA: Wellesley-Cambridge Press, 2009).

$$\approx 1.2480 \times 10^{-12} + 1.0000 \cdot \Delta x - 0.5000 \cdot (\Delta x)^2 + 0.3298 \cdot (\Delta x)^3$$
$$\approx 1.0000 \cdot \Delta x - 0.5000 \cdot (\Delta x)^2 + 0.3298 \cdot (\Delta x)^3, \tag{1.10}$$

which we claim we can use with insignificant numerical error. It is instructive to do a few calculations comparing a true evaluation of $\ln(1 + \Delta x)$ with the polynomial substitute and confirming for yourself the values of Δx for which this is really okay.

Armed with this new polynomial, we return to our original problem (Eq. 1.6) and write

$$\ln a = \frac{1}{\Delta t} \left(-\frac{\Delta t}{RC} - 0.5 \left(\frac{\Delta t}{RC} \right)^2 - 0.3298 \left(\frac{\Delta t}{RC} \right)^3 \right). \tag{1.11}$$

Now the full implications of $\Delta t \ll RC$ can be made clear. Since $\Delta t \ll RC$ is the same thing as saying $\Delta t / RC \ll 1$, we see that the terms of $\Delta t / RC$ of second order and higher in Eq. 1.11 diminish rapidly as we make Δt smaller. We can thus go even further in our approximation and neglect these terms, keeping firmly in mind that if the error this introduces bothers us, we can always make Δt smaller and smaller until the error does not bother us. Then we are left with

$$a = e^{-1/RC}, \tag{1.12}$$

which means, at long last, that the capacitor voltage evolves as

$$V_c(n \cdot \Delta t) = V_0 e^{-n \cdot \Delta t / RC}, \tag{1.13}$$

and we can finally write

$$V_c(t) = V_0 e^{-t/RC}. \tag{1.14}$$

This is the answer that we were expecting all along. What is important is how we got here. Based on physical reasoning, we came up with a discrete-time model for the system's behavior, and showed that solutions to the difference equations of this sort (see Eq. 1.4) have the form $(a^{\Delta t})^n$. We then solved for $a^{\Delta t}$, and finally explored the consequences of allowing Δt to become arbitrarily small compared to RC.

Mathematicians have an expression for our last step. They might say we "took the limit of Eq. 1.3 as Δt goes to zero." That is, we might have written Eq. 1.3 as

$$\lim_{\Delta t \to 0} \left(\frac{V_c(t + \Delta t) - V_c(t)}{\Delta t} + \frac{1}{RC} V_c(t) = 0 \right). \tag{1.15}$$

Of course, we now can appreciate that Δt does not go all the way to zero. It just gets arbitrarily small, such that the discrete "chunking" of time is indistinguishable from a continuous flow of time in whatever context is appropriate. Well, it turns out that the limit

$$\lim_{\Delta t \to 0} \frac{V_c(t + \Delta t) - V_c(t)}{\Delta t} \qquad (1.16)$$

occurs so often in the mathematics of continuous variables that we give ourselves an abbreviation, or a shorthand:

$$\lim_{\Delta t \to 0} \frac{V_c(t + \Delta t) - V_c(t)}{\Delta t} = \frac{d V_c(t)}{dt}. \qquad (1.17)$$

This shorthand is called the derivative, as in "the derivative of V_c with respect to t." You may or may not remember from when you first learned derivatives that Eq. 1.16 was the formal definition given to you. We may therefore rewrite Eq. 1.3 using the shorthand

$$\frac{d V_c(t)}{dt} + \frac{1}{RC} V_c(t) = 0. \qquad (1.18)$$

This is just a standard, first-order differential equation. The standard procedure here is to "guess" the solution Ae^{st}. Plugging this solution into Eq. 1.18 results in s being determined as $-1/RC$, and then we choose A to be V_0 in order to satisfy the initial conditions. The point of all this is that Eq. 1.18 does not spring out of a vacuum. Starting with Eq. 1.1, we took a very common-sense approach to solving a dynamical problem whose physics we understood pretty well. The approach represented by Eq. 1.18 takes for granted all of the insight that we gained by plodding through our discrete-time development. This is completely appropriate, as once the basics are understood it is important to streamline our methods as practical matter.

On a final note, we may interpret Eq. 1.18 in another way that makes its meaning jump off the page. We can write it as

$$\frac{d V_c(t)}{dt} = -\frac{1}{RC} V_c(t). \qquad (1.19)$$

Putting this equation into words, we might say "The rate of change of the voltage across the capacitor is proportional to the voltage across it at any given time, and inversely proportional to the value of the RC product. That rate of change has the opposite sign of the voltage across the capacitor at a given time, so the magnitude of the voltage is always decreasing. The system comes to rest, which is to say, the rate of change of the capacitor voltage goes to zero, only when the voltage across the capacitor itself is zero." We see that this differential equation is indeed a very natural way to describe the time evolution of an RC circuit.

1.1.2 Higher-Order Systems

It turns out that the discretized development of Section 1.1.1 is readily extensible to higher-order systems. The first thing to do is to figure out the equivalent of Eq. 1.17 for higher-order derivatives. It is helpful to introduce additional notation; we often write the first derivative of a function $f(t)$ with respect to time as $f'(t)$. That is,

$$f'(t) = \lim_{\Delta t \to 0} \frac{f(t + \Delta t) - f(t)}{\Delta t} = \frac{df(t)}{dt}. \tag{1.20}$$

Now we have a function $f'(t)$. We might ask, what is the time rate of change of this new function? You may remember that the answer is the second derivative of f with respect to t:

$$f''(t) = \frac{d}{dt} \frac{df(t)}{dt} = \frac{d^2 f}{dt^2}. \tag{1.21}$$

To figure out the equivalent of Eq. 1.17, we simply find the derivative of $f'(t)$,

$$f''(t) = \lim_{\Delta t \to 0} \frac{f'(t + \Delta t) - f'(t)}{\Delta t}, \tag{1.22}$$

and substitute the definition of $f'(t)$ from Eq. 1.20. Doing so yields

$$f''(t) = \lim_{\Delta t \to 0} \frac{f(t + 2 \cdot \Delta t) - 2f(t + \Delta t) + f(t)}{(\Delta t)^2}. \tag{1.23}$$

Repeating this procedure over and over again, we can get whatever order derivative we wish.

Higher-order derivatives come up quickly as we go beyond the complexity of the RC circuit in Figure 1.1. For example, consider the LC circuit in Figure 1.2. Proceeding in the same spirit that led to Eq. 1.1, we can write

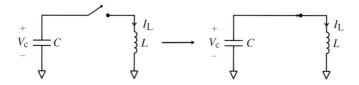

Figure 1.2 A simple LC circuit. The capacitor has an initial charge $Q_0 = CV_0$, and current begins to flow when the switch is closed at time $t = 0$.

$$CV_c(t + \Delta t) = CV_c(t) - I_L(t)\Delta t \tag{1.24}$$

$$I_L(t + \Delta t) = I_L(t) + \frac{1}{L}V_c(t)\Delta t.$$

One way to proceed from here is to solve the first equation for $I_L(t)$ in terms of $V_c(t)$ and $V_c(t + \Delta t)$, and then substitute for $I_L(t)$ and $I_L(t + \Delta t)$ in the second equation. Doing so causes our second-order derivative to appear right away:

$$\frac{V_c(t + 2 \cdot \Delta t) - 2V_c(t + \Delta t) + V_c(t)}{(\Delta t)^2} + \frac{1}{LC}V_c(t) = 0. \tag{1.25}$$

Now we proceed as before. We "guess" that $V_c(t)$ has the form $V_0 a^{n \cdot \Delta t}$, and are led to the quadratic characteristic equation in $a^{\Delta t}$:

$$(a^{\Delta t})^2 - 2(a^{\Delta t}) + 1 + \frac{(\Delta t)^2}{LC} = 0. \tag{1.26}$$

The quadratic formula readily provides us with possible values of $a^{\Delta t}$:

$$a^{\Delta t} = \frac{2 \pm \sqrt{4 - 4\left(1 + \frac{(\Delta t)^2}{LC}\right)}}{2} = 1 \pm j\frac{\Delta t}{\sqrt{LC}}. \tag{1.27}$$

As before, we take the log of both sides,

$$\ln a = \frac{1}{\Delta t}\ln\left(1 \pm j\frac{\Delta t}{\sqrt{LC}}\right), \tag{1.28}$$

only to encounter the log of a complex number. Dealing with this requires that we dust off a few important facts about complex numbers. The first is Euler's relation, which is

$$e^{j\theta} = \cos\theta + j\sin\theta. \tag{1.29}$$

The second fact is that any complex number $c + jd$ can be written in the polar form $re^{j\theta}$, where

$$r = \sqrt{c^2 + d^2} \tag{1.30}$$

and

$$\theta = \arctan\left(\frac{d}{c}\right). \tag{1.31}$$

The logarithm of this polar form is immediately apparent as

$$\ln(re^{j\theta}) = \ln r + j\theta. \tag{1.32}$$

Putting all of these facts together, we are free once again to pursue our original aim, which was solving for a. The argument of the logarithm in Eq. 1.28 becomes $re^{j\theta}$, where

$$r = \left(1 + \frac{(\Delta t)^2}{LC}\right)^{1/2} \tag{1.33}$$

and

$$\theta = \arctan\left(\frac{\Delta t}{\sqrt{LC}}\right). \tag{1.34}$$

So now Eq. 1.28 becomes

$$\ln a = \frac{1}{2 \cdot \Delta t} \ln\left(1 + \frac{(\Delta t)^2}{LC}\right) + \frac{j}{\Delta t} \arctan\left(\frac{\Delta t}{\sqrt{LC}}\right). \tag{1.35}$$

We now once again take a look at the consequences of a small Δt, this time noting that its smallness compared to $\sqrt{LC}$ is what counts. With the logarithmic term on the right side of Eq. 1.35, we do the same approximation that we did in Eq. 1.45. For the arctan term, we note that for $x \ll 1$, it can be shown that $\arctan x \approx x$. These approximations reduce Eq. 1.35 to

$$\ln a \approx \frac{1}{2} \frac{\Delta t}{LC} + \frac{j}{\sqrt{LC}}. \tag{1.36}$$

Here we notice one more thing: arbitrarily small Δt compared to $\sqrt{LC}$ makes for one further simplification, which is that

$$\ln a \approx \frac{j}{\sqrt{LC}}. \tag{1.37}$$

At the end of it all, we find that $a^{\Delta t} = e^{\pm j \cdot \Delta t / \sqrt{LC}}$, and therefore we can write the most general possible solution for V_c as

$$V_c(t) = A e^{+jt/\sqrt{LC}} + B e^{-jt/\sqrt{LC}}. \tag{1.38}$$

In actual applications, A and B are determined by the initial conditions for V_c and I_L. To see this, we can write the general solution for I_L using Eq. 1.24. Now that we are confident of its meaning, we freely employ the shorthand for the derivative and rewrite Eq. 1.24 as

$$I_L(t) = -C \frac{dV_c}{dt} = -\frac{j}{\sqrt{LC}} A e^{+jt/\sqrt{LC}} + \frac{j}{\sqrt{LC}} B e^{-jt/\sqrt{LC}}. \tag{1.39}$$

If we define the switch closing as $t = 0$, then the initial conditions dictate that

$$A + B = V_0 \qquad (1.40)$$

$$-\frac{j}{\sqrt{LC}}A + \frac{j}{\sqrt{LC}}B = 0.$$

With the aid of Euler's relation, it can be shown that capacitor voltage V_c and the inductor current I_L are given by

$$V_c = V_0 \cos(t/\sqrt{LC}) \qquad (1.41)$$

$$I_L = (V_0/\sqrt{LC})\sin(t/\sqrt{LC})).$$

As we leave this second-order example, it is worth reflecting on three things. First, the readers may feel some consternation over the use of complex numbers to describe the purely real variables V_c and I_L. It's not that we *chose* to use complex numbers; remember that they arose of algebraic necessity when solving Eq. 1.26. We'll have more to say about this in Section 1.2.3. Second, the oscillatory nature of the capacitor voltage is something that we might have guessed based purely on physical reasoning, before we did any math. For those readers who are comfortable with inductors and capacitors, it is worth going through the thought experiment at this point. Finally, notice that if we were not committed to a discretized development, we would have taken the limit as Δt goes to zero of Eq. 1.25 and written

$$\frac{d^2 V_c(t)}{dt^2} + \frac{1}{LC}V_c(t) = 0. \qquad (1.42)$$

The extension of these methods to higher-order systems proceeds in a straightforward way. Analysis typically leads to a system of first-order differential equations like Eqs. 1.19 and 1.24, which have the form[3]

$$\frac{dv_1}{dt} = a_{11}v_1 + \cdots + a_{1n}v_n, \qquad (1.43)$$

$$\frac{dv_2}{dt} = a_{21}v_1 + \cdots + a_{2n}v_n,$$

$$\vdots$$

$$\frac{dv_n}{dt} = a_{n1}v_1 + \cdots + a_{nn}v_n.$$

Ultimately we can collapse this system of equations down to one Nth-order differential equation in one of the "state variables"[4] v_i. Many readers have

---

[3] For much more on this, see the first chapter of William M. Siebert's excellent book *Circuits, Signals, and Systems* (Cambridge, MA: MIT Press, 1986).
[4] In our examples, the state variables have been V_c and I_L.

seen such developments in their sophomore-level electrical engineering classes on basic circuit analysis. What is hard to pick up the first time through this material is just how much differential equations really are just a common-sense, mathematical description of physical reality. The readers are encouraged to find a path to understanding differential equations at this intuitive level. This section, Section 1.1, is intended as a guide toward that end.

1.1.3 For Those of You Bothered by the Numerical Fitting in Section 1.1.1

A subset of you will wonder why we bothered with numerical fitting and didn't just use a Taylor expansion in place of Eq. 1.10. The answer has to do with the narrative: you can't dramatically unveil the derivative at the end of Section 1.1.1 if you've already employed it in a Taylor expansion.

But now we are past that. Recall that for a sufficiently well-behaved function $f(x)$, if we know its value at $f(x_0)$, we can write $f(x_0 + \Delta x)$ as the Taylor expansion:

$$f(x_0 + \Delta x) = f(x_0) + f'(x_0)\frac{\Delta x}{1!} + f''(x_0)\frac{(\Delta x)^2}{2!} + f'''(x_0)\frac{(\Delta x)^3}{3!} + \cdots$$
$$(1.44)$$

In theory a Taylor approximation has an infinite number of terms. When it comes to numerical computations, of course, we must truncate. The point is that we could have used the Taylor expansion to our advantage in Eq. 1.6, and avoided needing to go to a numerical aid like a computer. The natural log would be our f, and x_0 would be 1. Equation 1.6 would now become

$$\ln a = \frac{1}{\Delta t}\left(\ln(1) - \frac{\Delta t}{RC} - \frac{1}{2!}\left(\frac{\Delta t}{RC}\right)^2 - \frac{2}{3!}\left(\frac{\Delta t}{RC}\right)^3 + \cdots\right)$$
$$= \frac{1}{\Delta t}\left(-\frac{\Delta t}{RC} - \frac{1}{2}\left(\frac{\Delta t}{RC}\right)^2 - \frac{2}{3!}\left(\frac{\Delta t}{RC}\right)^3 + \cdots\right). \qquad (1.45)$$

Compare Eq. 1.45 to Eq. 1.11.

1.2 Convenient Properties of Linear Differential Equations

Much of classical control theory deals with linear differential equations with constant coefficients, two examples of which we examined in detail in Section 1.1. The general form for these equations is

$$A_n \frac{d^n y(t)}{dt^n} + A_{n-1} \frac{d^{n-1} y(t)}{dt^{n-1}} + \cdots + A_1 \frac{dy(t)}{dt} + A_0 y(t) \tag{1.46}$$

$$= B_m \frac{d^m x(t)}{dt^m} + B_{m-1} \frac{d^{m-1} x(t)}{dt^{m-1}} + \cdots + B_1 \frac{dx(t)}{dt} + B_0 x(t),$$

which we sometimes write as

$$\left(\sum_{i=0}^{n} A_i \frac{d^i}{dt^i} \right) y(t) = \left(\sum_{k=0}^{m} B_k \frac{d^k}{dt^k} \right) x(t). \tag{1.47}$$

Here $x(t)$ is understood to be an *input* or a *drive*, and $y(t)$ is the resultant output. It will be true of the systems that we are concerned with that $m \le n$.[5] We concern ourselves with equations of this form because they are solvable, because they have many nice mathematical properties, and because nature allows us to model an astonishing range of systems with this simple form.

1.2.1 Superposition!

The principle of *superposition*, or linearity, is one of the things we love about systems governed by linear differential equations with constant coefficients. Linearity is an extraordinarily important concept to grasp. Thankfully, the principle is easily and quickly stated. Suppose an operator A satisfies the relations

$$A(x_1(t)) = y_1(t) \tag{1.48}$$
$$A(x_2(t)) = y_2(t).$$

Now suppose we consider the superposition of our inputs, $\alpha x_1(t) + \beta x_2(t)$, where α and β are constants. It follows that if A is a linear operator, that is, *if superposition holds*, then it will be true that

$$A(\alpha x_1(t) + \beta x_2(t)) = \alpha y_1(t) + \beta y_2(t). \tag{1.49}$$

Another way to appreciate this concept is by considering a nonlinear system. We may take a squarer as an example, a system for which

$$A(x_1(t)) = x_1(t) \cdot x_1(t) \tag{1.50}$$
$$A(x_2(t)) = x_2(t) \cdot x_2(t).$$

In this case the results of superposing our inputs according to $\alpha x_1(t) + \beta x_2(t)$ would not be so easily separable:

---

[5] You can actually argue that this must be so based on our discretized development of Section 1.1. If $m > n$, the output at a particular time instant depends on the input at a *future* time instant. This is called a noncausal system, which is not of interest in real-time control applications.

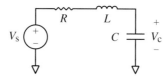

Figure 1.3 A series RLC network driven by a source $V_s(t)$.

$$A(\alpha x_1(t) + \beta x_2(t)) = \alpha^2 [x_1(t)]^2 + 2\alpha\beta \mathbf{x_1(t) \cdot x_2(t)} + \beta^2 [x_2(t)]^2. \quad (1.51)$$

Two aspects of this result leap off the page. The first is that when we scaled one of the inputs by α (or β), a portion of the output scaled as α^2 (or β^2). This is a satisfyingly nonlinear result, given that we specified the system to be a squarer. The second aspect, however, is the term in boldface, variously called an "intermodulation term," a "cross-modulation term," or a "mixing term." Such terms are completely typical of nonlinear systems, so much so that sometimes we explicitly exploit nonlinearity to build mixers and analog multipliers. The important point is that in linear systems, these intermodulation terms do not appear. We take full advantage of this fact when we introduce the "beautiful strategy" of Section 1.3.

To see how the abstract concept of linearity applies to real systems, consider the driven RLC circuit in Figure 1.3. A differential equation describing this system is

$$\frac{d^2 V_c(t)}{dt^2} - \frac{R}{L}\frac{dV_c(t)}{dt} - \frac{1}{LC}V_c(t) = -\frac{1}{LC}V_s(t). \quad (1.52)$$

The first thing to notice is that all of the operations done on $V_c(t)$ and $V_s(t)$ are linear operations. That is,

$$\frac{d^i}{dt^i}(\alpha v_1(t) + \beta v_2(t)) = \alpha\frac{d^i}{dt^i}v_1(t) + \beta\frac{d^i}{dt^i}v_2(t). \quad (1.53)$$

We are therefore assured that the relationship between the drive voltage $V_s(t)$ and the capacitor voltage $V_c(t)$ will be *linear*: if we double the amplitude of a particular drive voltage, for example, the corresponding capacitor voltage response will exactly double as well. Now suppose that we had tested the circuit by driving it with a drive $V_{s1}(t)$, and measured the capacitor voltage to be $V_{c1}(t)$. Suppose further that we follow that experiment with a second drive $V_{s2}(t)$, and the resultant capacitor voltage is $V_{c2}(t)$. Finally, suppose that we were curious what would happen if we drove the system with a superposition of the original two inputs, $\alpha V_{s1}(t) + \beta V_{s2}(t)$. Because this is a linear system (and time-invariant, as it happens, since the coefficients are constant), we would not need to do this third experiment. We would know right away that the capacitor voltage is the corresponding superposition of the solutions that we already determined by experiment, or $\alpha V_{c1}(t) + \beta V_{c2}(t)$. This is so important, and so fundamental, that it is vital that the readers *not* take the author's word for it. Work directly with Eq. 1.52, and convince yourself that the system described by this differential equation, or any differential equation with constant coefficients, is linear in the sense that we have described. Do not burden yourself with the need to provide a "proof" in the most formal sense, although if you have that skill so much the better. What is more important is that you get into the habit developing your own, satisfying explanations of things if the book that you are reading fails to deliver.

Before leaving this section, it is worth also formalizing what we mean by "time invariance." Returning our attention to Figure 1.3, suppose that we had driven the system with a specific input $V_s(t) = V_1(t)$, and determined that the resultant voltage was $V_c(t) = V_2(t)$. Now we ask, suppose we drive the circuit with the *delayed* input $V_1(t - t_0)$? There is a great temptation to overthink this. In any other context, you would say immediately that the output is just the same as it would have been before, only delayed by an amount t_0. That is, the capacitor voltage is just $V_2(t - t_0)$. You would be right! You can obtain some formal reassurance by solving Eq. 1.52 for the input e^{st}, and then solving it again for the input $e^{s(t-t_0)}$. Your "guess" in the second case will be that $V_c(t)$ is $Ae^{s(t-t_1)}$, and you will see that no solution for A is possible unless t_1 is equal to t_0. Since we will go on to argue that almost any input can be synthesized as a sum of complex exponentials (see Section 1.3.1), this little exercise is actually the start of a satisfyingly general demonstration.

The implications of linearity and time invariance for systems analysis are vast. All of the "frequency domain methods" that you have seen before, the usefulness of the Laplace, Fourier, and Z-transforms, the power of convolution sums and integrals, impulse responses (Green's functions in physics) – all of these analytical tools, and so many more, fall apart if the principle of superposition does not hold.

1.2.2 The Special Place of Exponentials

Each time we have jumped into the analysis of systems governed by linear differential equations with constant coefficients, as in Sections 1.1.1 and 1.1.2, complex exponentials have been waiting for us at the end. That is to say, in the undriven case our state variables have evolved according to a sum of terms like $C_i e^{st}$, where $s = \sigma + j\omega$, and where the C_i are determined by the initial conditions. We can now take a broader view of the special place of exponentials than simply that they pop up after a long and laborious process.

Consider again the general form of the differential equation

$$\left(\sum_{i=0}^{n} A_i \frac{d^i}{dt^i} \right) y(t) = \left(\sum_{k=0}^{m} B_k \frac{d^k}{dt^k} \right) x(t). \tag{1.54}$$

We will look at this for two cases. In the first we will consider the "zero-input response," or ZIR, which is how the system evolves when the drive $x(t) = 0$. Sections 1.1.1 and 1.1.2 were examples of ZIRs. The constants in ZIRs are chosen to satisfy all initial conditions. In the second case, we will consider the driven case assuming that all state variables have zero initial value, or the "zero-state response" (ZSR). It is worth emphasizing that splitting the solution into a ZIR and a ZSR is one of the many conveniences that linearity affords us. If $y_{ZIR}(t)$ is indeed a solution for the case of $x(t) = 0$, then it follows that

$$\left(\sum_{i=0}^{n} A_i \frac{d^i}{dt^i} \right) y_{ZIR}(t) = 0. \tag{1.55}$$

Suppose also that $y_{ZSR}(t)$ satisfies

$$\left(\sum_{i=0}^{n} A_i \frac{d^i}{dt^i} \right) y_{ZSR}(t) = \left(\sum_{k=0}^{m} B_k \frac{d^k}{dt^k} \right) x(t). \tag{1.56}$$

It necessarily follows that the sum $y_{ZIR}(t) + y_{ZSR}(t)$ also satisfies both Eq. 1.54 *and* the initial conditions for the system:

$$\left(\sum_{i=0}^{n} A_i \frac{d^i}{dt^i} \right) (y_{ZIR}(t) + y_{ZSR}(t)) = 0 + \left(\sum_{k=0}^{m} B_k \frac{d^k}{dt^k} \right) x(t)$$

$$= \left(\sum_{k=0}^{m} B_k \frac{d^k}{dt^k} \right) x(t). \tag{1.57}$$

It turns out that complex exponentials provide us with an easy path for determining both ZIRs and ZSRs.

The Zero-Input Response

As always with differential equations, we cannot claim to rigorously "solve" them. We must be content with "guessing" their solutions, and validating our guesses by demonstrating that the guessed solutions are, in fact, valid. Thankfully, linear differential equations with constant coefficients have long been sorted out: we know that the right guess in the undriven case ($x(t) = 0$)

is $y(t) = Ce^{st}$. The only question in this guess is what s and C are, exactly. We plug in to find out that

$$\left(\sum_{i=0}^{n} A_i \frac{d^i}{dt^i} \right) Ce^{st} = 0. \tag{1.58}$$

Differentiating exponentials is easy, and after dividing both sides by Ce^{st} we're left with n^{th}-order the polynomial in s:

$$A_n s^n + A_{n-1} s^{n-1} + \cdots + A_1 s + A_0 = 0. \tag{1.59}$$

And just like that, the introduction of the complex exponential reduces the unsolvable differential equation to a mere polynomial equation in the variable s, whose solution is available to anyone with access to a numerical solver. This is one reason why complex exponentials have a special place in linear system theory. Equation 1.59 is called the *characteristic equation* of the system. Solutions to the polynomial, or the *roots* of the characteristic equation, give us all of the values of s for which e^{st} is in fact a solution to the differential equation Eq. 1.58. And by linearity, if e^{st} is a solution then so is Ce^{st}. Linearity actually allows us to go further. For $n > 1$ there will be more than one value of s for which e^{st} is a solution to Eq. 1.58. In this case, the most general solution that we can write for Eq. 1.58 looks like[6]

$$y_{\text{ZIR}}(t) = C_n e^{s_n t} + C_{n-1} e^{s_{n-1} t} + \cdots + C_1 e^{s_1 t}. \tag{1.60}$$

Notice that *any* choice for the C_is results in a $y_{\text{ZIR}}(t)$ that satisfies Eq. 1.59. However, the initial conditions of the state variables determine these coefficients uniquely.

Finally, it should be emphasized that the s_i, the roots of the characteristic equation, are extremely special and particular to the system. If you "kick" the system, or by some means impose a set of initial conditions on the state variables, and then leave the system alone, all of the state variables will evolve in time according to the sum of a discrete set of exponentials. The only possibilities for the exponentials in this sum are those for which s satisfies the characteristic equation. These particular roots will come up again and again in our short review of linear systems in general and in our particular study of feedback systems.

---

[6] There are important departures when the characteristic equation does not have n distinct roots. See any number of textbooks on linear system theory for a fuller discussion, as well as Francis Hildebrand's *Advanced Calculus for Applications,* 2nd ed. (Englewood Cliffs, NJ: Prentice Hall, 1976).

The Zero-State Response

The ZSR is a little bit tougher to handle in general, and we'll postpone the more general treatment for just few sections more until Section 1.3. For now, however, the ZSR does give us yet another opportunity to extol the virtues of complex exponentials in the theory of linear systems.

In particular, suppose that we drive the system of Eq. 1.54 with the input $x(t) = e^{st}$, where as usual $s = \sigma + j\omega$. Continuing our highly successful strategy of correctly guessing the solution, in this case we posit $y(t) = Ce^{st}$ and move swiftly to validate:

$$\left(\sum_{i=0}^{n} A_i \frac{d^i}{dt^i}\right) Ce^{st} = \left(\sum_{k=0}^{m} B_k \frac{d^k}{dt^k}\right) e^{st}. \tag{1.61}$$

Derivatives of exponentials are easy, of course. With the special case of an exponential drive, the differential equation again collapses to polynomials in s:

$$(A_n s^n + A_{n-1} s^{n-1} + \cdots + A_0) C e^{st} = (B_m s^m + B_{m-1} s^{m-1} + \cdots + B_0) e^{st}. \tag{1.62}$$

The e^{st} factor appears in every term, and can be divided out. We are left with the condition on C that, when met, gives us the satisfaction of finding the true ZSR. The complex coefficient C is given by

$$C = \frac{B_m s^m + B_{m-1} s^{m-1} + \cdots + B_0}{A_n s^n + A_{n-1} s^{n-1} + \cdots + A_0}. \tag{1.63}$$

In Eq. 1.63 lies the validation of our initial guess. It affirms Ce^{st} as a solution by demonstrating that a C exists, and it determines C unambiguously as a function of s. If we make the dependence on s explicit, we might write Eq. 1.63 slightly differently:

$$C(s) = \frac{B_m s^m + B_{m-1} s^{m-1} + \cdots + B_0}{A_n s^n + A_{n-1} s^{n-1} + \cdots + A_0}. \tag{1.64}$$

In the theory of linear, time-invariant ("LTI" for short) systems, we call $C(s)$ a *transfer function*. If you know the transfer function for such a system, then you immediately know its response to the complex exponential $e^{s_0 t}$: it is just $C(s_0)e^{s_0 t}$. Because of linearity, you also know the response to $Ae^{s_0 t}$ is $AC(s_0)e^{s_0 t}$, and the response to $Ae^{s_0 t} + Be^{s_1 t}$ is $AC(s_0)e^{s_0 t} + BC(s_1)e^{s_1 t}$.

Notice that we see none of the cross-modulation that we saw in nonlinear systems (see Eq. 1.51). Evidently, the exponentials that you put in are the exponentials that you get out, only scaled by complex constants. Complex exponentials are sometimes called *eigenfunctions* of LTI systems for this reason,[7] and it is for this property that they occupy such a special place in LTI system theory.

At this point, you may suspect that there is more to this complex exponential stuff than presented so far. After all, strictly speaking e^{st} is a time function that exists for all time, and goes to infinity at either $t = +\infty$ or $t = -\infty$ if s has a nonzero real part. Since no physical input (or output, for that matter) ever goes to infinity, why is this eigenfunction business of any practical use? There are many ways that we deal with this issue within the theory of LTI systems.[8] However, what we will see shortly is that exponentials of the form $e^{j\omega t}$ are tremendously versatile and useful, and at least they don't blow up to infinity over any part of their domain.

1.2.3 But ... Why *Complex* Exponentials?

In brief: the reason we use complex exponentials is that they are convenient. We saw in Section 1.2.2 that derivatives and exponentials were made for each other. If we argue that differential equations are a natural way to express time evolution, then complex exponentials are far and away the easiest signals to work with when we are doing our math.

A concern that beginning engineering students have sometimes is that the imaginary parts of these complex exponentials are going to cause a practical problem. After all, in engineering one is using differential equations to model the voltages and currents in a circuit, or the displacement of a mass on a spring, or any other system in which the quantities are robustly and undeniably "real." It would not make sense in these contexts to do a calculation that insisted, say, that the final voltage across a capacitor was "$3j$ V."

A closer look at couple of earlier results will help to bolster your faith in the integrity of the LTI system theory machinery. First, look back at Section 1.1.2, where we had concluded that the most general solution for the capacitor voltage V_c was

$$V_c(t) = Ae^{+jt/\sqrt{LC}} + Be^{-jt/\sqrt{LC}}, \tag{1.65}$$

and the corresponding inductor current was

$$I_L(t) = -C\frac{dV_c}{dt} = -\frac{j}{\sqrt{LC}}Ae^{+jt/\sqrt{LC}} + \frac{j}{\sqrt{LC}}Be^{-jt/\sqrt{LC}}. \tag{1.66}$$

---

[7] In linear algebra, a matrix A will sometimes have certain special vectors $\vec{x}_i$ for which $A \cdot \vec{x}_i = a_i\vec{x}_i$, where a is a scalar. The matrix A does not alter the direction of special vectors $\vec{x}_i$ but only scales their lengths. The $\vec{x}_i$ are said to be *eigenvectors* of A.

[8] You may remember an insistence on using a *one-sided* Laplace transform, for example, or being very careful about "regions of convergence" in connection with the bilateral Laplace transform.

The coefficients A and B are determined by initial conditions $V_c(0) = V_0$ and $I_L(0) = I_0$, and for now let's consider the more general case of a parallel LC circuit for which the initial current needn't be zero. If we solve for A and B in terms of V_0 and I_0, we arrive at

$$A = \frac{1}{2}V_0 + \frac{1}{2}j\sqrt{LC}I_0 \qquad (1.67)$$

$$B = \frac{1}{2}V_0 - \frac{1}{2}j\sqrt{LC}I_0.$$

The key feature of this solution is that A and B are *complex conjugates* of each other. The real part of A equals the real part of B, while the imaginary part of A has the opposite sign of the imaginary part of B: $A = B^*$, or $A^* = B$. The fact that the initial conditions V_0 and I_0 are real guarantees this. If we now look back at Eqs. 1.65 and 1.66,

$$V_c(t) = Ae^{+jt/\sqrt{LC}} + A^*e^{-jt/\sqrt{LC}} \qquad (1.68)$$

$$I_L(t) = -\frac{j}{\sqrt{LC}}Ae^{+jt/\sqrt{LC}} + \frac{j}{\sqrt{LC}}A^*e^{-jt/\sqrt{LC}},$$

and expand using Euler's relation, what we will find is that $A = B^*$ means that $V_c(t)$ and $I_L(t)$ are purely real for all time!

It is beyond the scope of this book to provide a formal proof that this happy result will always occur. Nevertheless, the readers are encouraged to notice and appreciate the pattern. The values of s that satisfy our characteristic equations will always occur as either purely real roots, or complex conjugate pairs. For the ZIR, the coefficient of any $e^{(\sigma+j\omega)t}$ term will be the complex conjugate of the coefficient of the (always) accompanying $e^{(\sigma-j\omega)t}$ term. Therefore, for purely real initial conditions the ZIR will be real for all time.

The second reassuring result concerns the ZSR. Recall that if we excite an LTI system with a complex exponential e^{st}, the output will be $C(s)e^{st}$ with $C(s)$ given by Eq. 1.64:

$$C(s) = \frac{B_m s^m + B_{m-1}s^{m-1} + \cdots + B_0}{A_n s^n + A_{n-1}s^{n-1} + \cdots + A_0}. \qquad (1.69)$$

If we excite the system with a sinusoid $V_{in} = \sin(\omega_0 t + \phi)$, as we often do when testing electronic systems, we note that any purely real sinusoid can be written in terms of complex exponentials

$$\sin(\omega_0 t + \phi) = De^{j\omega_0 t} + D^*e^{-j\omega_0 t}, \qquad (1.70)$$

where, as it happens, $D = (-j/2)e^{j\phi}$. Exploiting linearity, we can immediately write down the response to this sinusoid using $C(s)$:

$$V_{\text{out}} = C(j\omega_0)De^{j\omega_0 t} + C(-j\omega_0)D^* e^{-j\omega_0 t}. \qquad (1.71)$$

We make one more important observation about $C(s)$: $C(s^*) = C^*(s)$. In LTI system theory, this is called being *conjugate symmetric*. Using this observation, we can rewrite Eq. 1.71 as

$$V_{\text{out}} = C(j\omega_0)De^{j\omega_0 t} + C^*(j\omega_0)D^* e^{-j\omega_0 t} \qquad (1.72)$$

and note with satisfaction that once again, the output voltage is purely real for all time. Apparently, if we input real sinusoids into LTI systems, we get real sinusoids out.

In summary, we use complex exponentials because they are extraordinarily convenient to work with. As long as the algebra is done correctly we needn't live in fear of a rogue imaginary part showing up and ruining the party.

Phasors Are a (Slightly) Different Trick

Use of phasors is another analytical technique that deserves mention in the context of complex exponentials. This is less because they are commonly used in feedback theory (they are not), and more because most engineering students have been exposed to phasors in their circuit analysis classes, electrodynamics classes, and in other instances. Confusion often results from not knowing when we are using phasors and when we are not. The gist of the phasor technique in circuit analysis is captured as follows: "Represent your sinusoidal drive as a the real part of a complex exponential; churn that complex exponential through whatever complex impedances you must; then drop the imaginary part at the end."

This is a different mode of operating than that represented in Eq. 1.70, in which we we expressed our sinusoidal drive literally as the sum of two complex sinusoids. By contrast, if we were using phasors we would have said that the input V_{in} is the *real part* of the complex quantity $-je^{j\phi}e^{j\omega_0 t}$, or $\text{Re}\{-je^{j\phi}e^{j\omega_0 t}\}$. The ZSR in response to this sinusoid still involves $C(s)$, but now our mathematical machinery does not automatically cancel the imaginary part for us. We must do it explicitly, and we do so by writing the output voltage is the real part of a complex expression:

$$V_{\text{out}} = \text{Re}\{-je^{j\phi}C(j\omega_0)e^{j\omega_0 t}\}. \qquad (1.73)$$

Fundamentally, the phasor method is based on what we know will ultimately happen: we *know* that the imaginary part will fall away anyway, so why not just drop it and save ourselves the fuss? It also brings clarity to realize that "dropping the imaginary" part can be easily mechanized. For a complex quantity A, we can get the real part of A as follows:

$$\text{Re}\{A\} = (1/2)A + (1/2)A^*. \tag{1.74}$$

If we apply this to Eq. 1.73, we get

$$V_{\text{out}} = (-j/2)e^{j\phi}C(j\omega_0)e^{j\omega_0 t} + (j/2)e^{-j\phi}C(-j\omega_0)e^{-j\omega_0 t}. \tag{1.75}$$

We see that this is exactly the same as Eq. 1.72.

The point of this section is that the readers needn't maintain a boundary in their minds between phasors on one hand and the rest of linear system theory on the other. The phasor method is a very clever exploitation of the conjugate symmetry of expressions like $C(s)$, which in turn inherit their remarkable properties from the natural reasoning process that gives us differential equations.

1.3 Frequency Domain Methods: A Beautiful Strategy

Now we get to arguably the most important section of this chapter. Frequency domain methods and thinking permeate almost all of engineering and physics. There is perhaps no more important concept to get a firm grounding in for a student entering these fields.

For purposes of this book, this section answers a huge question that was hinted at in Section 1.2.2. We saw that finding the zero-state response of a system in response to a complex exponential is especially easy for systems that are well described by linear differential equations with constant coefficients. The great unanswered question: so what?! In the real world, one does not encounter exponential drives that started at the beginning of time and have been rising exponentially ever since! Nor does one encounter sinusoids that continue for all time. What use is this great edifice of LTI systems and mathematical tricks to an engineer if we cannot use them to gain insight into physical systems?

The remarkable answer is that a broad class of signals of can be represented as a weighted sum of complex exponentials. This is true even of signals of finite duration. And remember that for the systems of interest to us, superposition holds (see Sec. 1.2.1). These two facts suggest the following beautiful strategy for analyzing the response of LTI systems in response to *arbitrary* inputs:

1. Break the input up into a sum of complex exponentials.
2. Pass each of these exponentials individually through the LTI system. Each exponential will result in a new exponential scaled in amplitude and phase shifted according to an expression like $C(s)$ in Eq. 1.63. These are the output exponentials.
3. Sum up the output exponentials. This is now the output signal, and the response to the input of step 1.

Many readers will not remember the strategy this way. Instead, what may come to mind is a feverish rush through three different-looking steps:

1. Take the Fourier (or Laplace) transform of the input signal to get $X(j\omega)$!
2. Multiply $X(j\omega)$ by the system transfer function $H(j\omega)$!
3. Look up the inverse transform of $H(j\omega)X(j\omega)$ to get $y(t)$!

It turns out that the two lists of steps are the same. The purpose of the remaining subsections of Section 1.3 are to illustrate how.

1.3.1 Fourier Series Representation of Periodic Signals

We will start with the first step, "Break the input up into a sum of complex sinusoids." This is the meaning and purpose of the Fourier transform, and a good way to understand the Fourier transform is to first understand the Fourier series representation of periodic signals.

It turns out that almost any periodic signal can be represented as a sum of sinusoids. Suppose that the signal in question $x(t)$ has period T. That is to say, regardless of how funny a shape $x(t)$ has, it is always true that $x(t+T) = x(t)$. It is a remarkable fact that almost all such signals can be represented as a sum of sinusoids like

$$x(t) = A_0 + A_1 \cos(\frac{2\pi}{T}t + \phi_1) + A_2 \cos(2 \cdot \frac{2\pi}{T}t + \phi_2) \cdots, \qquad (1.76)$$

where the A_is and the ϕ_is are determined by the detailed shape of $x(t)$. This is a fact that is actually familiar from everyday experience. If you hear a specific pitch played on a violin, a trumpet, and sung by a singer, it will be obvious that all three musicians are producing the same "note." That is, something internal to each instrument is vibrating at the same number of repetitions per second. In the language of engineering physics, we say that the signal produced by each musician has the same fundamental period T. It will be equally obvious that although the note is the same, there is a world of difference between a note played on the violin versus played on a trumpet, and no one will confuse the sound of a trumpet or a violin with that of the human voice. In music this is called a difference in *timbre*, and a musician may speak of the difference in "overtones," "harmonics," or "partials"[9] produced by the three musicians. If we were to put a recording of these musicians through a spectrum analyzer, what we would find is that the different timbres would manifest as different ratios A_i/A_1s and ϕ_is for $i \geq 2$ for each instrument.

We can express $x(t)$ more compactly as

$$x(t) = \sum_{k=0}^{\infty} A_k \cos(k\omega_0 t + \phi_k), \qquad (1.77)$$

where $\omega_0 = 2\pi/T$. Our challenge is now to determine what the A_i actually are, but before we do so it is useful to recast Eq. 1.76 in terms of complex exponentials. Noting that

$$A_k \cos(k\omega_0 + \phi_k) = \frac{1}{2}A_k e^{j\phi_k} e^{jk\omega_0 t} + \frac{1}{2}A_k e^{-j\phi_k} e^{-jk\omega_0 t}, \qquad (1.78)$$

we can let C_k be a complex constant defined as $C_k = \frac{1}{2}A_k e^{j\phi_k}$, and also let $C_{-k} = \frac{1}{2}A_k e^{-j\phi_k}$. Doing this allows us to rewrite Eq. 1.77 in terms of complex exponentials:

$$x(t) = C_0 + (C_1 e^{j\omega_0 t} + C_{-1} e^{-j\omega_0 t}) \qquad (1.79)$$
$$+ (C_2 e^{j2\omega_0 t} + C_{-2} e^{-j2\omega_0 t}) \cdots$$
$$= \sum_{k=-\infty}^{\infty} C_k e^{jk\omega_0 t}.$$

Now, the interesting part: how to determine the C_ks, the individual coefficients, if we know the shape of $x(t)$? Consider the following:

$$\int_{t_0}^{t_0+T} e^{jk\omega_0 t} e^{-jl\omega_0 t}\, dt = \int_{t_0}^{t_0+T} e^{j(k-l)\omega_0 t}\, dt \qquad (1.80)$$

$$= \begin{cases} \frac{1}{j(k-l)\omega_0}[e^{j(k-l)\omega_0 t}]_{t_0}^{t_0+T} & k \neq l \\ T & k = l \end{cases}$$

$$= \begin{cases} 0 & k \neq l \\ T & k = l. \end{cases}$$

---

[9] In music, a concert "A" has a fundamental frequency of 440 Hz, or a fundamental period T of 2.27 ms. An octave up is the second harmonic, which is also an "A" at 880 Hz and is sometimes called the "second partial" or "first overtone." At 1320 Hz is the third harmonic/partial and second overtone, etc.

If we accept that periodic signal $x(t)$ can be represented as the *Fourier series* in Eq. 1.79, we can use Eq. 1.80 to determine a formula for the C_ks:

$$\int_{t_0}^{t_0+T} x(t)e^{-jl\omega_0 t}\,dt = \int_{t_0}^{t_0+T}\left(\sum_{n=-\infty}^{\infty} C_k e^{jk\omega_0 t}\right)e^{-jl\omega_0 t}\,dt \qquad (1.81)$$

$$= TC_l.$$

From here, it follows that

$$C_l = \frac{1}{T}\int_{t_0}^{t_0+T} x(t)e^{-jl\omega_0 t}\,dt. \qquad (1.82)$$

This is a critical point in the development. We started with the proposition that almost any periodic signal could be represented as a sum of complex sinusoids, and came up with a recipe for determining exactly which sinusoids, and how much of each sinusoid, belongs in the sum. This is important enough that it is worth summarizing the two equations together. The first is the "synthesis" equation, so called because it shows how the original function can be synthesized by adding together its sinusoidal components. The second is the "analysis" equation, which shows how the original signal can be analyzed by breaking it into its constituent sinusoids:

Synthesis equation:

$$x(t) = \sum_{k=-\infty}^{\infty} C_k e^{jk\omega_0 t} \qquad (1.83a)$$

Analysis equation:

$$C_l = \frac{1}{T}\int_{t_0}^{t_0+T} x(t)e^{-jl\omega_0 t}\,dt. \qquad (1.83b)$$

With the synthesis and analysis equations in hand, a powerful strategy for dealing with arbitrary periodic input $x(t)$ to LTI systems starts to emerge. Using the analysis equation, we can break the $x(t)$ up into a sum of complex sinusoids $x(t) = \sum_{k=-\infty}^{\infty} C_k e^{jk\omega_0 t}$.[10] Then we can rely on superposition to know that that the output $y(t)$ will be a superposition of the *exact same set of sinuosids*. That is, the output $y(t)$ is given by $y(t) = \sum_{k=-\infty}^{\infty} D_k e^{jk\omega_0 t}$. An excellent use of Eq. 1.64 is to determine the D_k explicitly:

---

[10] A concern arises: computationally, how would one deal with the fact that this is technically an infinite sum? The answer is that we would approximate the input by limiting $|k| \le K$, where K is chosen large enough to keep the error to an acceptable level.

$$D_k = C_k \frac{B_m \cdot (jk\omega_0)^m + B_{m-1} \cdot (jk\omega_0)^{m-1} + \cdots + B_0}{A_n \cdot (jk\omega_0)^n + A_{n-1} \cdot (jk\omega_0)^{n-1} + \cdots + A_0}. \tag{1.84}$$

The importance of Eq. 1.84 is that it is a first example of analyzing system behavior in the *frequency domain*. In Eq. 1.84 we have made use of a transfer function that we first encountered in Eq. 1.64. The transfer function is a mapping of the frequency components of the input to the frequency components of the output, and is given to us naturally by the differential equation. By way of contrast, up until now we have considered the problem of system analysis in terms of finding a *time-domain* description of the behavior of the output. At least for periodic signals, we have discovered a very powerful mode of analysis. For aperiodic signals, we need the Fourier transform.

1.3.2 The Fourier Transform and the Meaning of Integrals

We emerge from Section 1.3.1 with a powerful pair of relations for hopping back and forth between the time and frequency domains, repeated here for convenience:

$$x(t) = \sum_{k=-\infty}^{\infty} C_k e^{jk\omega_0 t} \tag{1.85}$$

$$C_l = \frac{1}{T} \int_{t_0}^{t_0+T} x(t) e^{-jl\omega_0 t} dt.$$

This is great for periodic waveforms, but what about aperiodic waveforms? It turns out we can point our way to the answer if we consider a different question: what would we do with a waveform that was periodic, but whose fundamental period T is one million years? In terms of our lifetimes, or indeed the lifetimes of any human civilization, there is no practical distinction between a signal with a period of a million years or a truly aperiodic signal. So perhaps it makes sense to think of aperiodic signals in terms of Eq. 1.85 in the limit as T gets very large.

One thing that happens when T gets large is that the steps between frequency components get very small. The size of the step is

$$\Delta \omega = \omega_0 = \frac{2\pi}{T}. \tag{1.86}$$

A first step on the way to the Fourier transform can be to change the first of our relationship pair to be in terms of $\Delta \omega$:

$$x(t) = \sum_{k=-\infty}^{\infty} C_k e^{j(k \cdot \Delta \omega)t}. \tag{1.87}$$

We will also make one more change, which is to redefine C_l according to

$$C_l = \int_{t_0}^{t_0+T} x(t)e^{-jl\omega_0 t}\,dt, \tag{1.88}$$

so that $x(t)$ is now given by

$$x(t) = \frac{1}{T} \sum_{k=-\infty}^{\infty} C_k e^{jk\cdot\Delta\omega t}. \tag{1.89}$$

Noting that $\frac{1}{T} = \frac{\Delta\omega}{2\pi}$, we arrive at the slightly different-looking pair:

$$x(t) = \frac{1}{2\pi} \sum_{k=-\infty}^{\infty} C_k e^{j(k\cdot\Delta\omega)t}\,\Delta\omega \tag{1.90}$$

$$C_l = \int_{-T/2}^{T/2} x(t)e^{-j(l\cdot\Delta\omega)t}\,dt.$$

Looking at $x(t)$, we now consider what happens in the limit of $T \to \infty$. Formally this is written

$$x(t) = \lim_{T\to\infty} \frac{1}{2\pi} \sum_{k=-\infty}^{\infty} C_k e^{j(k\cdot\Delta\omega)t}\,\Delta\omega \tag{1.91}$$

$$= \lim_{\Delta\omega\to 0} \frac{1}{2\pi} \sum_{k=-\infty}^{\infty} C_k e^{j(k\cdot\Delta\omega)t}\,\Delta\omega,$$

which looks an awful lot like the limit of a Riemann sum that we have come to know as an integral. For suitably well-behaved $x(t)$, this is exactly how we can treat Eq. 1.91. The limit becomes

$$x(t) = \frac{1}{2\pi} \int_{-\infty}^{\infty} C(\omega)e^{j\omega t}\,d\omega. \tag{1.92}$$

We speak of $C(\omega)$ as the *Fourier transform* of $x(t)$, and the convention is to express C as a function of $j\omega$ and not merely ω, and to label it X instead of C. Following this convention, the synthesis/analysis equation pair is now

$$x(t) = \frac{1}{2\pi} \int_{-\infty}^{\infty} X(j\omega)e^{j\omega t}\,d\omega \tag{1.93}$$

$$X(j\omega) = \int_{-\infty}^{\infty} x(t)e^{-j\omega t}\,dt.$$

$X(j\omega)$ is sometimes called a *frequency domain* description of the signal $x(t)$.

A major point here is that the integral, like the derivative, is nothing particularly fancy, despite the enduring impression drilled into most engineering undergraduates. We just get tired of writing

$$\lim_{\Delta x \to 0}$$

all over the place and give ourselves a $\int$ or a $\frac{d}{dt}$ as a shorthand. It is pretty much that simple.

It is worth internalizing this understanding of the integral and the derivative. It will help you to have a unified view of continuous-time and discrete-time systems, and also help to understand how computers cope with "continuous" variables.

1.3.3 The Strategy

Now it is appropriate to revisit the opening of Section 1.3. The goal is to figure out how an LTI system responds to an arbitrary input. The first step is

Break the input up into a sum of complex exponentials.

By "break into a sum of exponentials," we now know that we mean find $X(jk \cdot \Delta \omega)$ for all k such that

$$x(t) = \lim_{\Delta \omega \to 0} \frac{1}{2\pi} \sum_{k=-\infty}^{\infty} X(jk \cdot \Delta \omega) e^{j(k \cdot \Delta \omega)t} \Delta \omega$$

is an equality. The idea is that for each value of k, $X(jk \cdot \Delta \omega)$ tells us how heavily to weight $e^{j(k \cdot \Delta \omega)t}$ in the sum of exponentials. Moving to the continuous limit, we write that

$$X(j\omega) = \int_{-\infty}^{\infty} x(t) e^{-j\omega t} dt \qquad (1.94)$$

and congratulate ourselves for finishing the first step.

The next step is

Pass each of these exponentials individually through the LTI system.
Each exponential will result in a new exponential scaled in amplitude
and phase shifted according to an expression like C(s) in Eq. 1.63.
These are the output exponentials.

We know that complex exponentials are special to LTI systems: they pass through peacefully, and have only their amplitude and phase modified. We can write a general relationship, then, between the sinusoids that compose $x(t)$ and the sinusoids that compose the output $y(t)$. Relying on Eq. 1.64, we have

$$Y(j\omega) = X(j\omega)\frac{B_m(j\omega)^m + B_{m-1}(j\omega)^{m-1} + \cdots + B_0}{A_n(j\omega)^n + A_{n-1}(j\omega)^{n-1} + \cdots + A_0}.$$

We'll do one more thing, which is give the transfer function its own symbol $H(j\omega)$, so that the frequency-domain representation of our output is now

$$Y(j\omega) = X(j\omega)H(j\omega).$$

The last step is as follows:

Sum up the output exponentials. This is now the output signal, and the response to the input of step 1.

And now we see, at long last, the expression that you doubtless remember from many, many problems as an undergraduate. Summing up the output exponentials means

$$y(t) = \frac{1}{2\pi}\int_{-\infty}^{\infty} Y(j\omega)e^{j\omega t}\,d\omega = \frac{1}{2\pi}\int_{-\infty}^{\infty} X(j\omega)H(j\omega)e^{j\omega t}\,d\omega.$$

Note that you almost never actually did this integral. Instead, you learned to recognize certain forms of $X(j\omega)H(j\omega)$ and relied on a table of inverse Fourier transforms to get your $y(t)$s. What you see now is the underlying logic of the whole thing.

1.4 Impulses in Linear, Time-Invariant Systems

As neatly as the story falls together, any discussion of continuous functions that are "broken up into their infinitely close constituent parts" is bound to have some awkward moments. In this section we discuss the *impulse function*, which helps us to get through these.

1.4.1 Why Impulses?

Our first awkward moment comes when we consider the Fourier transform of $x(t) = e^{j\omega_0 t}$, the vaunted complex exponential that nature so favors. Diving in,

$$X(j\omega) = \int_{-\infty}^{\infty} e^{j\omega_0 t}e^{-j\omega t}\,dt, \qquad (1.95)$$

we discover to our dismay that for $\omega \neq \omega_0$, the integral is zero, while for $\omega = \omega_0$ the integral blows up! What in the world just happened? Our "beautiful" framework seems to have cost us the ability to represent a simple sinusoid!

Notice, though, that when we get back to our summation roots, there is no problem. Recall that

$$x(t) = \lim_{\Delta\omega \to 0} \frac{1}{2\pi} \sum_{k=-\infty}^{\infty} X(jk \cdot \Delta\omega) e^{j(k \cdot \Delta\omega)t} \Delta\omega. \qquad (1.96)$$

All is right if we specify $X(\cdot)$ according to

$$X(jk \cdot \Delta\omega) = \begin{cases} \frac{j2\pi}{\Delta\omega} & k \cdot \Delta\omega = \omega_0 \\ 0 & k \cdot \Delta\omega \neq \omega_0, \end{cases} \qquad (1.97)$$

where the assumption is that $\Delta\omega$ is chosen such that there is exactly one integer k for which $k \cdot \Delta\omega = \omega_0$.

Now as we go to the continuous limit of $\Delta\omega \to 0$, we seem to making extraordinary demands of the function $X(j\omega)$. A frequency-domain representation of a sinusoid seems to involve a function that goes to infinity over an infinitesimally small span, and then just vanishes everywhere else. It is enough to drive our mathematician friends nuts.

We needn't fret. It brings a fresh dose of perspective to recognize that even for supposedly continuous distributions, there is a resolution limit beneath which we are helpless to look. Consider, for example, the difference between a sinusoid at 10 Hz and one at $(10 + 10^{-12}$ Hz). If you beat these two sinusoids against one another, you get a beat once every 30,000 years. It is very difficult, as a practical matter, to tell these two sinusoids apart. And if one argues that they have built a spectrum analyzier that *does* have this kind of extraordinary resolution, we can repeat the example with a difference of 10^{-18} Hz, or however extreme we need to be to get the point across.

It remains that this line of discussion leaves the more analytical reader cold. It is not a good feeling to have mathematical analysis tied somehow to measurement limits. The short story of how this issue is resolved is that we give ourselves a notational "out" called the impulse or δ function. We gaze upon it with only the soft focus required to define its behavior under an integral:

$$\int_{-\infty}^{\infty} \delta(\omega) d\omega = 1 \qquad (1.98)$$

$$\int_{-\infty}^{\infty} \delta(\omega - \omega_0) X(j\omega) d\omega = X(j\omega_0).$$

Resist the urge to define the impulse in any more detail than this. Ruminations on the exact shape of the impulse functions are particularly unprofitable.

It is enough that the impulse function rescues us from the awkwardness of discussing infinite numbers of infinitesimally spaced entities.

1.4.2 The Fourier Transform and the Impulse Response

We saw in Section 1.4.1 that the impulse function comes in handy for frequency-domain descriptions of signals. It turns out that time-domain impulses, $\delta(t)$, are useful as well. Figure 1.4 shows three different voltage

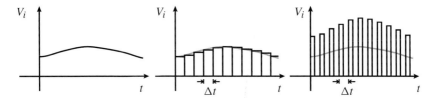

Figure 1.4 For a sufficiently slow-responding system, these three inputs are equivalent.

waveforms as the input to an unspecified system. The first input is a smoothly varying voltage. The second approximates the first as series of rectangular pulses. The third is also a series of rectangular pulses, but the pulses are twice as tall and half as long as those in the second input so the *area* of the pulses is the same. It turns out that if Δt is sufficiently small compared to the response time of the driven system, all three of these inputs will produce nearly identical outputs. And if we do break the input up into pulses, for sufficiently small Δt the shape of the pulses is not critical. Only the area of the pulses matters; we draw rectangular pulses here because rectangles are easy to draw.[11]

You can probably see where we are headed with this: only the area of the pulses matter; the duration has to be extremely "short" relative to any physical time scale; we *must* be talking about impulses. Indeed, having already seen how we can break up a signal into a sum of sinusoids, we find that we can also break up a signal in the time domain into a *superposition of time-delayed impulses*. Such a superposition looks like this:

---

[11] In his book *Circuits, Signals, and Systems*, William M. Siebert describes the "smoothing effect of physical systems." This smoothing effect may be worth exploring on your own. A series RC circuit driven by a voltage source (as shown in Figure 1.5, for example), is a fine place to start playing with pencil and paper. Use as a drive rectangular pulses of duration significantly shorter (10x or more) than the time constant $\tau = RC$. With some work, you will start to see strong dependence on the area of the pulse, but a relative insensitivity to the details of a pulse's height and duration.

$$x(t) \Rightarrow \sum_n x(n \cdot \Delta t) \Delta t \cdot \delta(t - n \cdot \Delta t). \tag{1.99}$$

It is worth reading Eq. 1.99 carefully. We are exploiting the "smoothing effect of physical systems" to substitute for the true $x(t)$ a series of time-delayed impulses. Since we assign area of each impulse in our sum to be $x(n \cdot \Delta t) \Delta t$, it is evident that we are thinking of the rectangular pulse representation in Figure 1.4. We use the area of each rectangular pulse to scale an impulse, whose area would otherwise be 1 (see Eq. 1.98). For purposes of analysis we do not dwell on the exact value of Δt, other than to assert that it is sufficiently small relative to the response time of the system under discussion so as to justify the substitution in Eq. 1.99.

Now we are in position to powerfully leverage the properties of linearity and time invariance that we saw in Section 1.2.1. Suppose that we had characterized the system in terms of its *impulse response* $h(t)$. That is to say, in response to a unit impulse $\delta(t)$, the output of the system is $h(t)$. Linearity and time invariance dictate that in response to a scaled and shifted impulse $A\delta(t - t_0)$, the response of the system would be $Ah(t - t_0)$. Applying this reasoning to the general input $x(t)$ of Eq. 1.99, we can write the output $y(t)$ of an LTI system in terms of its impulse response as

$$y(t) = \sum_n x(n \cdot \Delta t) \Delta t \cdot h(t - n \cdot \Delta t). \tag{1.100}$$

This is called a *convolution sum*, and in the context of continuous-time systems we are accustomed to taking the limit as Δt goes to zero and writing the *convolution integral*,

$$y(t) = \int_{-\infty}^{\infty} x(\tau) h(t - \tau) d\tau, \tag{1.101}$$

which winds up being the same as

$$y(t) = \int_{-\infty}^{\infty} h(\tau) x(t - \tau) d\tau. \tag{1.102}$$

We often skip the integral notation and write this as

$$y(t) = x(t) * h(t) = h(t) * x(t). \tag{1.103}$$

The impulse response is useful as a way of characterizing the extent to which a system is or is not "memoryless." Suppose, for example, that we have an extremely fast system for which the response to an impulse is just an impulse, $h(t) = \delta(t)$. Remembering the defining properties of the impulse in Eq. 1.98, the result of applying Eq. 1.101 is that $y(t) = x(t)$. That is, the output at any given instant depends only on the input at that *same* instant. That is said to be a "memoryless" system. Slower-responding systems, or systems that act as low-pass filters, will tend to have impulse responses that are extended in time instead of being highly localized, and their outputs at any given instant will depend both on the current input and on the history of prior inputs.

Finally, we can tie our discussion of the impulse response to the discussion of transfer functions that emerged in Section 1.3.3. Recalling that the first step of the Strategy is to break the input up into a sum of complex exponentials, we take the Fourier transform of our input, which is now $x(t) = \delta(t)$:

$$X(j\omega) = \int_{-\infty}^{\infty} \delta(t)e^{-j\omega t}\,dt = 1. \qquad (1.104)$$

Apparently, this remarkable entity requires sinusoids of *all* frequencies to synthesize it!

The next step is to pass all of these sinusoids through the system to see how they are affected. We did this with the aid of the transfer function $H(j\omega)$, and write the Fourier transform of the output $y(t)$ as

$$Y(j\omega) = X(j\omega)H(j\omega) = 1 \cdot H(j\omega) = H(j\omega). \qquad (1.105)$$

This means that the transfer function $H(j\omega)$ is nothing other than the Fourier transform of the impulse response! This is a very important result and worth internalizing. The impulse response $h(t)$ is the time-domain characterization of a system, $H(j\omega)$ is a frequency-domain characterization (sometimes called its "frequency response"), and $h(t)$ and $H(j\omega)$ form a Fourier transform pair.

1.4.3 The Fourier Transform of Differential Equations

The Fourier transform can be viewed as a tool for dealing with differential equations. As we have seen time and again, one cannot simply derive the solution to a differential equation. We must content ourselves with guessing the proper form of the solution, and plugging in and verifying that it is in fact valid. But with the Fourier transform in hand, we can turn a differential equation into an algebraic equation. And those we can solve.

To see how this arises, consider the Fourier transform of the first-order derivative of $y(t)$:

$$\int_{-\infty}^{\infty} \frac{dy(t)}{dt} e^{-j\omega t}\,dt.$$

It is easiest to integrate by parts, after agreeing to a couple of convenient preconditions. The first is that $y(t)$ starts at time $t = 0$, and is zero for all time before that. Certainly for the real-time control systems that concern this book, we can do this without loss of generality. The second precondition is that $y(t)$ is of finite extent, so that $y(\infty) = 0$. Agreeing to these, we proceed to integrate by parts:

$$\int_{-\infty}^{\infty} \frac{dy(t)}{dt} e^{-j\omega t} dt = \int_{0}^{\infty} \frac{dy(t)}{dt} e^{-j\omega t} dt \qquad (1.106)$$

$$= \int_{0}^{\infty} \frac{d}{dt}\left[y(t)e^{-j\omega t}\right]dt - \int_{0}^{\infty} y(t)\left[\frac{d}{dt}e^{-j\omega t}\right]dt$$

$$= -y(0) + j\omega \int_{0}^{\infty} y(t)e^{-j\omega t} dt$$

$$= j\omega \cdot Y(j\omega) - y(0).$$

We will normally be concerned with the ZSR, for which $y(t)$ and all of its derivatives are zero at $t = 0$. Equation 1.106 becomes

$$\int_{-\infty}^{\infty} \frac{dy(t)}{dt} e^{-j\omega t} dt = j\omega \cdot Y(j\omega). \qquad (1.107)$$

Moreover, if you investigate the Fourier transform of higher-order derivatives of $y(t)$, what you find is that

$$\int_{-\infty}^{\infty} \frac{d^n y(t)}{dt^n} e^{-j\omega t} dt = (j\omega)^n \cdot Y(j\omega). \qquad (1.108)$$

Now consider again a general form of a differential equation that describes a dynamic system:

$$\left(\sum_{i=0}^{n} A_i \frac{d^i}{dt^i}\right) y(t) = \left(\sum_{k=0}^{m} B_k \frac{d^k}{dt^k}\right) x(t). \qquad (1.109)$$

If there is equality between the left and right sides of this equation, it must be true that the Fourier transform of the left side is equal to the Fourier transform of the right. Utilizing Eq. 1.107, the *differential* equation Eq. 1.109 becomes the *algebraic* equation

$$\left(\sum_{i=0}^{n} A_i \cdot (j\omega)^i\right) Y(j\omega) = \left(\sum_{k=0}^{m} B_k \cdot (j\omega)^k\right) X(j\omega). \qquad (1.110)$$

While we could not properly solve Eq. 1.109, Eq. 1.110 is easy. $Y(j\omega)$ is given simply by

$$Y(j\omega) = \frac{\sum_{k=0}^{m} B_k \cdot (j\omega)^k}{\sum_{i=0}^{n} A_i \cdot (j\omega)^i} X(j\omega). \qquad (1.111)$$

If Eq. 1.111 looks familiar, it is because we saw it before in Eq. 1.64. The readers are encouraged to reconcile how exactly it is that we got to the same place through two different-looking routes.

1.5 The Unilateral Laplace Transform

We are just about finished developing tools for LTI system analysis. We have one more method to introduce, which is the unilateral Laplace transform. Even though we have saved it for last here, it is the most widely used "transform method" used in analyzing LTI systems. Whereas with the Fourier transform we probed deeply into its conceptual meaning, the Laplace transform is best seen as a rather impersonal "crank" to turn. It is simply a method to take a differential equation and turn it into an algebraic equation. The unilateral Laplace transform $X(s)$ of a time-domain waveform $x(t)$ is

$$X(s) = \int_{0_-}^{\infty} x(t)e^{-st}dt. \tag{1.112}$$

We choose the lower limit as "0_-" to avoid ambiguity when $x(t)$ contains an impulse at the origin. It is understood that the *entire* impulse is to be counted under the integral.[12] Also, note that for real-time control applications such as we are concerned with in this book, choosing a lower limit of "0_-" as opposed to "$-\infty$" is convenient, as it is equivalent to assuming that all of our signals start no earlier than $t = 0$.

The inverse Laplace transform is slightly more complicated, involving a complex contour integral. The proper definition is

$$x(t) = \frac{1}{2\pi j} \int_{\sigma_0-j\infty}^{\sigma_0+\infty} X(s)e^{st}ds. \tag{1.113}$$

One way to handle integrals of this form is to use the calculus of residues.[13] In practice, however, what we find is that the $X(s)$s for which we need to do this integral come in relatively few distinct forms. It is usually more efficient to use a table of inverse transforms, which is the reason you probably do not recall ever using the calculus of residues. Tables of inverse Laplace transforms can be found in any textbook on linear system theory, and a short one is provided at the end of this chapter.

---

[12] The issue of the proper lower limit for a unilateral Laplace transform gets an extensive treatment in Thomas Kailath's book *Linear Systems* (Harlow, UK: Prentice Hall, 1979).

[13] A thorough discussion of this technique can be found in Hildebrand's *Advanced Calculus for Applications,* 2nd ed. (Englewood Cliffs, NJ: Prentice Hall, 1976).

Once we have the definitions of the Laplace transform and its inverse, we apply them to differential equations in the same way that we did in Section 1.4.3. Consider the now-familiar general form of a differential equation:

$$\left(\sum_{i=0}^{n} A_i \frac{d^i}{dt^i}\right) y(t) = \left(\sum_{k=0}^{m} B_k \frac{d^k}{dt^k}\right) x(t). \tag{1.114}$$

Taking the Laplace transform of both sides, we once again turn a differential equation into an algebraic equation, this time in the variable s:

$$\left(\sum_{i=0}^{n} A_i \cdot s^i\right) Y(s) = \left(\sum_{k=0}^{m} B_k \cdot s^k\right) X(s). \tag{1.115}$$

Now the solution for $Y(s)$ is simply

$$Y(s) = \frac{\sum_{k=0}^{m} B_k s^k}{\sum_{i=0}^{n} A_i s^i} X(s). \tag{1.116}$$

One of the nice consequences of using the Laplace transform, as opposed to the Fourier transform, is that the link to characteristic equation (like Eq. 1.59) is made typographically explicit.

1.5.1 Dynamic Interpretation of Poles

The poles of a transfer function figure prominently in any discussion of an LTI system. Before we get into a discussion of them, remember that what drives us in analysis is, in fact, the need to boil incredibly complex systems down to a bare minimum of information or characteristics. The "minimum" is of course context dependent. Are we trying to make a design decision? Are we curious about the time scale on which the system will respond? Or are we instead concerned about the fate of every individual charge carrier in the system? Assuming that we are not interested in that last (hopelessly low) level of detail, it turns out that the pole locations contain some very important information about the speed of response of a given system.

Take the circuit of Figure 1.5 as a simple but useful example upon which we will wield our newfound tool, the unilateral Laplace transform. As is now our custom, we can derive a differential equation that describes the time evolution of, say, the voltage across the capacitor C. That equation is

$$RC \frac{dV_c(t)}{dt} + V_c(t) = V(t). \tag{1.117}$$

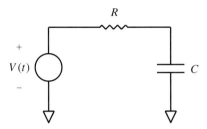

Figure 1.5 A driven RC circuit.

Now, however, we take the Laplace transform of both sides. We then write the relationship between $V_c(s)$ and $V(s)$ as the straightforward

$$V_c(s) = \frac{1/RC}{s + 1/RC} V(s). \qquad (1.118)$$

It is customary to define a transfer function $H(s)$ connecting $V_c(s)$ and $V(s)$:

$$H(s) = \frac{1/RC}{s + 1/RC}. \qquad (1.119)$$

$H(s)$ is a function of the complex variable $s = \sigma + j\omega$. This function has a *singularity* at $s = -1/RC$. That is to say, as s approaches the special value $-1/RC$, the magnitude of $H(s)$ goes to infinity. In the theory of complex variables, this kind of singularity is called a *pole*.

Putting that aside for the moment, we return to our characterization of the system in Eq. 1.118 and ask, what is the output in response to an impulse? Since the Laplace transform of an impulse is simply unity for all values of s, the Laplace transform of the response is given by

$$V_c(s) = \frac{1/RC}{s + 1/RC}. \qquad (1.120)$$

Now we follow the script and perform the inverse transform,

$$V(t) = \frac{1}{2\pi j} \int_{\sigma - j\infty}^{\sigma + \infty} \left(\frac{1/RC}{s + 1/RC} \right) e^{st} ds, \qquad (1.121)$$

with, it goes without saying, the aid of a table of inverse Laplace transforms:

$$V(t) = \frac{1}{RC} e^{-t/RC} u(t). \qquad (1.122)$$

We define $u(t)$, the unit step function, according to

$$u(t) = \begin{cases} 1 & t \geq 0 \\ 0 & t < 0. \end{cases} \qquad (1.123)$$

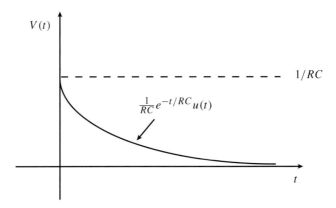

Figure 1.6 The response of a series RC circuit to a voltage impulse.

Graphically, the impulse response is shown in Figure 1.6. In response to an impulse, the voltage across the capacitor jumps up abruptly, and then decays exponentially to zero. The *time constant*, or the amount of time that must pass for the voltage across the capacitor to decay to $1/e$ times its initial value, is given by the product RC. But RC is the reciprocal of the pole location of the transfer function given in Eq. 1.120. And here is the key, the critical thing about pole locations: *the locations of the poles tell us how quickly the system will settle to its final value after an excitation.*

This is an important enough concept that it is worth dwelling upon briefly. Suppose that instead of exciting the system with an impulse, we instead excite the system with a step $u(t)$. Taking the Laplace transform of the unit step, which is $1/s$, and multiplying by the transfer function gives us the Laplace transform of the output

$$V_c(s) = \frac{1/RC}{s(s + 1/RC)}. \tag{1.124}$$

To use the inverse transform tables easily, we first rewrite this as[14]

$$V_c(s) = \frac{1}{s} - \frac{1}{(s + 1/RC)}. \tag{1.125}$$

The time-domain $V_c(t)$ is given by

$$V_c(t) = (1 - e^{-t/RC})u(t). \tag{1.126}$$

---

[14] The tool that we use here is called a partial fraction expansion, and can be found in almost all textbooks on linear system theory. A good, inexpensive one to own is Zoher Karu's *Signals and Systems Made Ridiculously Simple* (Cambridge, MA: ZiZi Press, 1995).

This is the *step* response, and is shown in Figure 1.7. The final value of this response is 1 V. In this case, notice that the difference between the final value and the value of the transient at a given time fades as $e^{-t/RC}$.

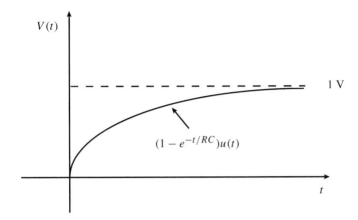

Figure 1.7 The response of a series RC circuit to a unit voltage step. The voltage across the capacitor asymptotically approaches 1 V.

As a final example to illustrate the importance of poles, consider the impulse response of a transfer function that has multiple poles:

$$H(s) = \frac{K_0(s - z_1)(s - z_2) \cdots (s - z_m)}{(s - p_1)(s - p_2) \cdots (s - p_n)}. \tag{1.127}$$

In this example we make reference to the general transfer function of Eq. 1.116, where the numerator may also be a polynomial in s whose roots are called the *zeros* of the transfer function. Again utilizing a partial fraction expansion, and recognizing that $n \geq m$, what emerges is the impulse response $h(t)$:[15]

$$h(t) = A_1 e^{p_1 t} + A_2 e^{p_2 t} + \cdots + A_n e^{p_n t}. \tag{1.128}$$

So in response to an impulse, each pole accounts for a portion of the settling behavior in a very clear, transparent way. Poles with a real part greater than zero will give to the impulse response a term that "blows up," or rises exponentially. We therefore consider systems with such "right-half-plane" (RHP) poles to be unstable. For the fast poles, or poles for which $Re\{p_i\} < 0$ and $|Re\{p_i\}|$ is large compared to that of the other poles, their contribution to the response will fade to insignificance very quickly. Indeed, the slow poles, or poles for which $|Re\{p_i\}|$ is small compared to the other poles, will tend

---

[15] This form assumes no repeated roots in the numerator or denominator.

to dominate the response. For this reason, we tend to call the slowest pole in the system the "dominant" pole, as the time response tends to be well approximated by a single-pole system whose single pole is the dominant pole of the original system. It is worth taking the time to make sure that you really see this.

It is also worth briefly circling back to the discussion of Figure 1.4. At that point, there was a vague assertion that for a "sufficiently slow-responding system," we could replace our continuous driving signal with a series of short pulses or impulses spaced Δt apart in time. Once we know the poles of the system, we can be more precise about the size of Δt be relative to the time constant of the dominant pole in the system. Figure 1.8 shows that when Δt is small compared to $1/|p_{dominant}|$, where $p_{dominant}$ is the dominant pole in the system, we are working at a time scale for which we needn't differentiate between a continuous drive and a drive with scaled and shifted impulses. This insight is particularly important in time-domain computer simulators for dynamic systems, where discretization of the time variable is inescapable. Designers of easy-to-use simulators devote substantial effort to automatic determination of a time step that is small enough to ensure accuracy, while not being so small that the computation becomes impractical.

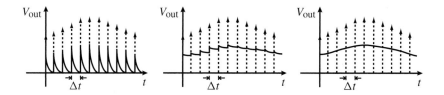

Figure 1.8 Response of a system to an impulse train when $\Delta t \gg \tau$, $\Delta t \approx \tau$, and $\Delta t \ll \tau$, where τ is the time constant of the dominant pole in the system.

1.5.2 The Geometric View of Poles and Zeros

An understanding of poles and zeros of transfer functions is foundational to a study of feedback systems. The purpose of this section is to remind readers of the geometric view of these singularities. If this topic is new to readers, it may be helpful to first browse through the root locus discussion of Section 2.5. Root locus techniques are a major application of the geometric picture of poles and zeros for feedback systems. This section also provides important grounding for the discussion in Chapter 3 of the Nyquist stability criterion.

Consider again the generic transfer function,

$$H(s) = \frac{K_0(s - z_1)(s - z_2)\cdots(s - z_m)}{(s - p_1)(s - p_2)\cdots(s - p_n)}, \tag{1.129}$$

and focus on one term in the denominator:

$$(s - p_n). \tag{1.130}$$

In this term, s is a complex variable, and p_n is a complex constant. For now we will make their complex nature explicit by letting $s = s_r + js_i$, and $p_n = p_{rn} + p_{in}$ and rewriting Eq. 1.130 as

$$(s - p_n) = (s_r - p_{rn}) + j(s_i - p_{in}). \tag{1.131}$$

So far, nothing remarkable has happened. We have just stated some obvious, arithmetic facts, and rewritten our original statement in somewhat more detail.

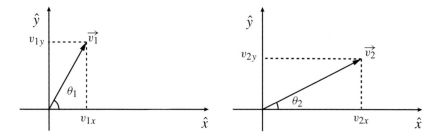

Figure 1.9 Examples of vectors familiar from geometry.

This expanded form of $s - p_n$, is familiar. Recall vectors from geometry:

$$\vec{v_1} = v_{1x}\hat{x} + v_{1y}\hat{y} \tag{1.132}$$
$$\vec{v_2} = v_{2x}\hat{x} + v_{2y}\hat{y}.$$

We would always draw these vectors as shown in Figure 1.9. It turns out that $\theta_1 = \arctan(v_{1y}/v_{1x})$, $\theta_2 = \arctan(v_{2y}/v_{2x})$. Now when we wanted to do the vector subtraction $\vec{v_1} - \vec{v_2}$ *graphically*, we started by rotating $\vec{v_2}$ by 180°. We then translated the rotated $\vec{v_2}$ such that its tail met the head of $\vec{v_1}$. The result of the vector subtraction is then given graphically by drawing an arrow from the tail of $\vec{v_1}$ to the head of the inverted $\vec{v_2}$. Pictures are obviously better than words here; see Figure 1.10. Looking closely at this geometric procedure, what we see is that it is the visualization of

$$\vec{v_1} - \vec{v_2} = (v_{1x} - v_{2x})\hat{x} + (v_{1y} - v_{2y})\hat{y}. \tag{1.133}$$

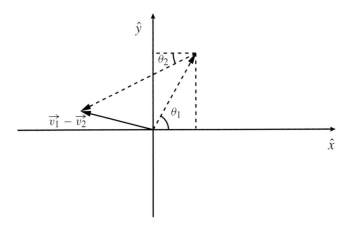

Figure 1.10 Graphical vector subtraction.

Now let's look back at terms like $(s - p_n)$. There is a striking similarity between the subtraction of complex numbers in Eq. 1.131 and the subtraction of vectors shown in Eq. 1.133. That similarity is that the equations are the same but for two simple substitutions, with 1 taking the role of $\hat{x}$ and j taking the role of $\hat{y}$. It will turn out to be incredibly useful to exploit this similarity by viewing complex numbers as vectors in a "complex plane." That is, $a + jb$ becomes as shown in Figure 1.11. The *real* axis takes the place of the $\hat{x}$ axis, and the *imaginary* axis takes the place of the $\hat{y}$ axis. Notice that this simple mapping of $\hat{x}$ to 1 and $\hat{y}$ to j is actually the second hint we've been given to see complex numbers as vectors in the complex plane. The first was provided by Euler's relation, which we can use to write a complex number as

$$a + jb = a\cos\theta + j \cdot a\sin\theta \qquad (1.134)$$

$$= re^{j\theta}.$$

Looking at Figure 1.11, it is obvious that $r = \sqrt{a^2 + b^2}$ and $\theta = \arctan(b/a)$.

Where are we going with this? Once we accept the geometrical picture of complex numbers, it turns out that there is a graphical way to represent how a complex sinusoid will be affected by passing through a transfer function with a given set of poles and zeros. Consider the individual terms from Eq. 1.129, with those in the numerator described by $r_{zi}e^{j\theta_{zi}} = (s - z_i)$ and those in the denominator described by $r_{pi}e^{\theta_{pi}} = (s - p_i)$. If we pass a complex sinusoid $e^{st} = e^{\sigma t}e^{j\omega t}$ through a system described by the transfer function in Eq. 1.129, we substitute $s = \sigma + j\omega$ and we find that this sinusoid will be *scaled* and *phase shifted*. The new amplitude of the sinusoid will be

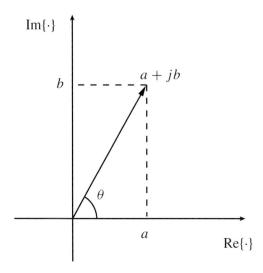

Figure 1.11 A complex number as a vector.

$$r_o = \frac{r_{z1} r_{z2} \cdots r_{zm}}{r_{p1} r_{p2} \cdots r_{pn}}, \tag{1.135}$$

and it will be phase shifted according to

$$\theta_o = \theta_{z1} + \theta_{z2} + \cdots + \theta_{zm} - \theta_{p1} + \theta_{p2} + \cdots + \theta_{pn}, \tag{1.136}$$

so that what emerges from the system is the scaled and shifted sinusoid

$$y(t) = r_o e^{\sigma t} e^{j(\omega t + \theta_o)}. \tag{1.137}$$

Now for the graphical way of seeing how sinusoids are affected by poles and zeros. Consider the individual denominator term $(s - p_i)$, and the input sinusoid $e^{j\omega_0 t}$. For this input, $s = j\omega_0$. Vectors illustrating the pole p_i and the input sinusoid are shown in Figures 1.12 and 1.13.

To get some practice with the graphical method, it is extremely helpful to sketch out simple cases for which you already know what to expect from prior experience with Bode plots. For example, a pole at the origin will give a phase shift of $-90°$ for all nondecaying sinusoids $e^{j\omega t}$. A real pole in the left-half plane will give a negligible phase shift for sinusoids near DC; when $|p| = |\omega|$, the phase shift imparted will be $-45°$ and the sinusoid will be scaled by $1/\sqrt{2}$ (or, famously, -3dB!); and when $|p| \ll |\omega|$, the sinusoid becomes almost completely attenuated and the phase shift goes asymptotically to $-90°$. Verify these statements using the graphical method, and then try making up a few cases of your own. How does a lightly damped pole pair affect $e^{j\omega t}$ for various values of ω?

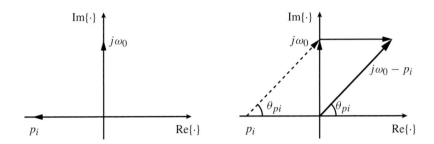

Figure 1.12 Graphical evaluation of how sinusoid $e^{j\omega_0 t}$ is affected by a pole. The length of $j\omega_0 - p_i$ indicates how the sinusoid is scaled, and the angle θ_{pi} indicates how it is phase shifted.

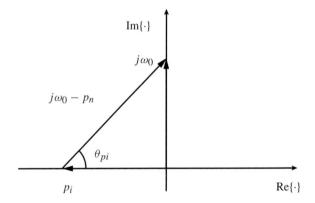

Figure 1.13 A redrawing of Figure 1.12, skipping intermediate steps. To see graphically how pole p_i affects sinusoid $e^{j\omega_0 t}$, the pole and the complex frequency s of the sinusoid are each placed in the complex plane. A vector is then drawn from the pole to s. The sinusoid is scaled by the reciprocal of the length of the vector, and the phase shift is **minus** θ_{pi}.

The insight that can be gleaned from this graphical representation of poles and zeros will be leaned on quite heavily when we turn our attention to root locus and Nyquist plots. Give this section careful consideration now, and if it is new to you, come back to it as you work through Chapters 2 and 3.

1.5.3 Initial and Final Value Theorems

The initial value theorem and the final value theorem should be in your linear system theory bag of tricks. We use the final value theorem in particular for feedback systems to answer the question: "What is the steady-state error of my controller in response to a step input?"[16]

The final value theorem concerns the relationship between the final value of a time domain signal $x(t)$ and its Laplace transform $X(s)$. In the language of mathematics, we render the final value of $x(t)$ as

$$\lim_{t \to \infty} x(t). \tag{1.138}$$

The final value theorem gives us the equivalence

$$\lim_{t \to \infty} x(t) = \lim_{s \to 0} s X(s). \tag{1.139}$$

To show that this is true, a good place to start is by writing out $s X(s)$ in terms of $x(t)$:

$$\lim_{s \to 0} s X(s) = \lim_{s \to 0} s \int_{0_-}^{\infty} x(t) e^{-st} dt \tag{1.140}$$

$$= \lim_{s \to 0} \int_{0_-}^{\infty} x(t) \left[s e^{-st} \right] dt$$

$$= \lim_{s \to 0} \int_{0_-}^{\infty} x(t) \frac{d}{dt} \left[-e^{-st} \right] dt.$$

To integrate by parts, recognize that

$$x(t) \frac{d}{dt} \left[e^{-st} \right] dt = \frac{d}{dt} \left[x(t) e^{-st} \right] - \frac{dx(t)}{dt} e^{-st}. \tag{1.141}$$

We continue from Eq. 1.140 according to

$$= \lim_{s \to 0} \left[-\int_{0_-}^{\infty} \frac{d}{dt} \left[x(t) e^{-st} \right] dt + \int_{0_-}^{\infty} \frac{dx(t)}{dt} e^{-st} dt \right] \tag{1.142}$$

$$= \lim_{s \to 0} \left[-\left[x(t) e^{-st} \right]_{0_-}^{\infty} + \int_{0_-}^{\infty} \frac{dx(t)}{dt} e^{-st} dt \right].$$

From here, we must be explicit about how we take the limit on s. For the one-sided signals for which the unilateral Laplace transform makes sense, the Laplace transform integral itself safely converges if $\text{Re}\{s\} > 0$. Therefore, we take the limit $s \to 0$ with s approaching zero from the right half of the complex plane. Under these circumstances, the proper evaluation of the limit in s is

---

[16] The step input is important because it represents commanding the system to go to and hold a specific value. For an example, in the cruise control of a car a step input would be commanding the car to go to 55 mph and stay at that speed. The steady-state error is the difference between 55 mph and the speed the car ultimately settles to. The ideal is for the steady-state error to be zero.

$$= -\left[0 - x(0_-)\right] + [x(t)]_{0_-}^{\infty} \tag{1.143}$$

$$= \lim_{t \to \infty} x(t).$$

The initial value theorem has a satisfyingly similar form:

$$\lim_{t \to 0} x(t) = \lim_{s \to \infty} s X(s). \tag{1.144}$$

This relationship can be shown using steps very similar to those used to establish the final value theorem.

1.5.4 Inverting the Laplace Transform

Interestingly, inversion of the Laplace and Fourier transforms is typically the one skill that a desperately time-constrained, scrambling, first-time student of linear systems is able to acquire. For a chapter that concerns what the reader "may have missed the first time through," it is therefore fitting to only include a table of common transforms. For a review of and practice with inverting common Laplace transforms, as well as more comprehensive tables, see books by Siebert and by Karu, cited in full in Chapter 6. Meantime, a short table of Laplace transforms is given in Table 1.1.

Table 1.1 *A short table of Laplace transform pairs*

x(t)		X(s)
$\delta(t)$	$\Longleftrightarrow$	1
$\delta(t - T)$	$\Longleftrightarrow$	e^{-sT}
$u(t)$	$\Longleftrightarrow$	$\frac{1}{s}$
$e^{-\alpha t}$	$\Longleftrightarrow$	$\frac{1}{s+\alpha}$
t^n	$\Longleftrightarrow$	$\frac{n!}{s^{n+1}}$
$\sin \omega_0 t$	$\Longleftrightarrow$	$\frac{\omega_0}{s^2+\omega_0^2}$
$\cos \omega_0 t$	$\Longleftrightarrow$	$\frac{s}{s^2+\omega_0^2}$

1.6 Convolution and the Special Place of Exponentials

Back in Section 1.4.2, we "discovered" the convolution integral while exploring the impulse function. For a linear, time-invariant system driven characterized by impulse response $h(t)$, the response $y(t)$ to input $x(t)$ is given by

$$y(t) = x(t) * h(t) = \int_{-\infty}^{\infty} x(\tau)h(t-\tau)d\tau = \int_{-\infty}^{\infty} h(\tau)x(t-\tau)d\tau.$$
(1.145)

Notice what happens now if choose for the input the complex exponential $x(t) = e^{s_0 t}$:

$$y(t) = \int_{-\infty}^{\infty} h(\tau)x(t-\tau)d\tau \qquad (1.146)$$

$$= \int_{-\infty}^{\infty} h(\tau)e^{s_0(t-\tau)}d\tau$$

$$= e^{s_0 t}\int_{-\infty}^{\infty} h(\tau)e^{-s_0 \tau}d\tau$$

$$= H(s_0)e^{s_0 t}.$$

Through examination of the convolution integral in this way, it becomes evident that the eigenfunction nature of complex exponentials is a direct consequence of linearity and time invariance.

1.7 Discrete-Time Formalism: Same Ideas, Different Notation

The mathematics for discrete-time and continuous-time systems are typically taught as separate but related topics in undergraduate linear systems courses. The connection between the two modes of thought is usually left implicit. Students are aware, for example, that their intuitive understanding transfers easily between the two types of systems even if the notation seems wildly different. Once students have thoroughly digested both discrete-time and continuous-time analysis, it is a good time for them to explore the relationship between the two explicitly.

1.7.1 Difference Equations Are a *Really* Natural Expression of Time Evolution

This book starts by asserting that differential equations are a natural way to express time evolution. The reason that discussion ultimately got complicated is that the time step, Δt, must approach the limit of zero if we are to truly represent continuous time. This inconvenient fact spawned a derivative- and integral-heavy notation that can obscure the common sense of what is happening.

We are spared such acrobatics when we turn our attention to discrete-time systems. Our attention is on an output $y[n]$, and we simply desire to know what happens next. That is, what is $y[n + 1]$? To be as general as we can, we must assume that $y[n + 1]$ is influenced by present and past values of the input $x[n]$. And if we allow for the possibility of feedback, then there ought to be a dependence on present and past values of $y[n]$ as well. It bears emphasizing that "time" in discrete-time systems is divorced from the everyday concept of time. We would use the same notation whether the system was stepping in increments of years or in nanoseconds. Think about and internalize that.

Mathematically, we want to capture the dependency of $y[n + 1]$, what happens *next*, on present and past inputs and outputs. The most general way to do this is by writing

$$y[n + 1] = \sum_{k=0}^{\infty} a_k x[n - k] + \sum_{j=0}^{\infty} b_j y[n - j]. \qquad (1.147)$$

If the notation seems obscure after all of the integrals we have been writing, go back to the very first example considered in this book. The first equation that we wrote, Eq. 1.1, was

$$Q_c(t + \Delta t) = Q_c(t) - I(t)\Delta t. \qquad (1.148)$$

We made a guess to the solution of this equation involving $a^{n \cdot \Delta t}$, and plugged it into Eq. 1.4. Ultimately, we explored the implications of Δt being very small in order to arrive at a continuous-time solution. It was the pursuit of that limit that led to the introduction of the derivative, as well as the ubiquitous appearance of e^{st} in our solutions. In the mathematics of discrete time, there is no Δt to take to a limit. We simply write

$$Q_c[n + 1] = Q_c[n] - I[n] \qquad (1.149)$$

and pursue a development that closely parallels that of Section 1.1. By doing this development yourself, you can clarify in your mind the underlying similarity between continuous-time and discrete-time methods.

1.7.2 The Fourier Transform in Discrete Time

It is not difficult to see that difference equations, like their differential equation cousins, have a soft spot in their hearts for complex exponentials. Suppose that the input to a system described by Eq. 1.147 is given by $x[n] = z_0^n$, where $z_0 = re^{j\Omega}$. As with differential equations, we posit a guess for $y[n]$ and see if it is consistent with the governing equation. Taking a cue from the differential equation development, we might suppose that the output is none other than the input scaled by a complex constant, or $y[n] = Az_0^n$. So much for the guess; now we plug into Eq. 1.147 and follow our nose:

$$y[n+1] = \sum_{k=0}^{\infty} a_k x[n-k] + \sum_{j=0}^{\infty} b_j y[n-j], \qquad (1.150)$$

$$Az_0^{n+1} = \sum_{k=0}^{\infty} a_k z_0^{n-k} + A\sum_{j=0}^{\infty} b_j z_0^{n-j},$$

$$A = \frac{\sum_{k=0}^{\infty} a_k z_0^{n-k}}{z_0^{n+1} - \sum_{j=0}^{\infty} b_j z_0^{n-j}}.$$

Equation 1.150 is the discrete-time cousin of the continuous-time transfer function Eq. 1.64.

The Fourier transform for discrete-time analysis is developed in a method very similar to that of the continuous-time case. The readers are encouraged to pursue this development independently, either by reading in the references or deriving it themselves. The result will be the synthesis and analysis equivalents of the continuous-time relations (Eq. 1.93):

$$x[n] = \frac{1}{2\pi} \int_{-\pi}^{\pi} X(e^{j\Omega}) e^{j\Omega n} d\Omega \qquad (1.151)$$

$$X(e^{j\Omega}) = \sum_{n=-\infty}^{\infty} x[n] e^{-j\Omega n}.$$

Notice that there are some important differences, however. First, for the synthesis formula we do not integrate over an infinite domain. Instead, the limits of integration are from $-\pi$ to π. Fundamentally, this arises from the fact that the time index is restricted to integers: we find that for all integers k, $e^{(\Omega+2\pi k)n}$ is indistinguishable from $e^{\Omega n}$. This will become clearer when we talk about aliasing in Section 4.3.1.

More confusing sometimes is the *form* of the complex exponentials that we deal with in discrete vs. continuous time. In continuous time we talked of complex exponentials e^{st}, where s is the complex frequency $\sigma + j\omega$. However, in discrete time our exponentials are z^n, where z itself is the complex number that we concern ourselves with. The difference is superficial, as can be seen by recognizing that the complex number z can always be written $re^{j\Omega}$. But if it is superficial, why bother? We bother because the representations e^{st} and z^n reflect what is easy to work with in differential equations and difference equations, respectively. Sometimes, yes, it comes down to that.

1.7.3 The Z-Transform, the Impulse Response, and Convolution in Discrete Time

The parallels to continuous-time analysis will keep coming. In place of the unilateral Laplace transform, we have the unilateral Z-transform:

$$X(z) = \sum_{\infty}^{\infty} x[n]z^{-n}. \tag{1.152}$$

The inverse Z-transform, with all of the usual caveats about how it is rarely used, involves the contour integral

$$x[n] = \frac{1}{2\pi j} \oint_C X(z)z^{n-1}dz. \tag{1.153}$$

Impulses in discrete time come with none of the baggage that weighs down impulses in continuous time. A discrete-time impulse is given simply by

$$\delta[n] = \begin{cases} 1 & n = 0 \\ 0 & n \neq 0. \end{cases} \tag{1.154}$$

Finally, looking back it can be seen that the argument leading to convolution in Section 1.4.2 is much easier to make in discrete time. For a discrete-time system characterized by impulse response $h[n]$, the response to arbitrary input $x[n]$ is given by

$$y[n] = \sum_m x[m]h[n-m]. \tag{1.155}$$

It is also true that the Z-transform of the impulse response in discrete time is equal to the transfer function. For tables of common Z-transform pairs, see Siebert's excellent book on signals and systems, cited in full in Chapter 6.

1.8 Chapter Summary

If there is one underlying theme in this chapter, it is that linear system theory *makes sense*. At the end of the day, it is nothing more than common sense written in the language of mathematics. Do not be fooled by fancy differential equations, integrals, summations, and transforms. They are a shorthand that, once you achieve fluency, you too will find convenient.

One other point bears emphasis as we bring this chapter to a close. Part of the reason that we love linear system theory is that its equations can usually be solved. Even better, the solutions to those equations yield insight that is powerful for design. These aspects of linear system theory perhaps cannot be properly appreciated without studying the theory of nonlinear systems, which are often far more complicated and the analysis of which often yields far less insight. In fact, a common strategy for analyzing nonlinear systems is to approximate them as linear systems and apply the methods we have described here. It turns out that a good understanding of linear systems will give readers a solid basis for understanding many types of physical systems in the real world.

2

The Basics of Feedback

Feedback is more than just something that we use every day. For us, it is a mode of being. It is worth taking a look at examples of feedback in everyday life to clarify what we mean by a "closed-loop" (feedback) versus a "open-loop" (nonfeedback) system. This exercise can greatly sharpen one's grasp of the conceptual framework of feedback systems. Bear in mind, however, that the actual processes utilized by a human being are far, far more sophisticated than the feedback systems that can be covered by the theory presented in this book. The fundamental difference between a "feedback" process undertaken by a human being and one undertaken by a dumb servo loop is that humans (and animals) use past experience to optimize the performance of a task. By contrast, an analog feedback loop can be hit by a unit step a million times in a row, and a million times in a row it will overshoot by the same amount, ring the same amount, and always take the same amount of time to settle to its final value. So while the following discussion gets across the spirit of feedback, readers should understand that an op-amp or a servo controller is a is a mere toy compared to what we in the animal kingdom regularly deploy.

2.1 Filling a Glass with Water

The first challenge we consider is filling a glass of water. In this scenario, the "command" is that we fill the glass up until it is "full," but not so close to the top that it is awkward to handle. Since we are getting exact, we might specify that our definition of "full" is that the glass is filled until the water level is *exactly* 0.5 inches from the top of the glass, as shown in Figure 2.1.

Now we all know how we, personally, would go about filling the glass. This was a major challenge for us as toddlers, but as adults it has long since

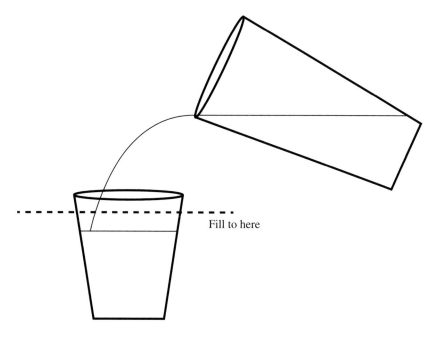

Fill to here

Figure 2.1 Filling a water glass.

moved past the point of being a skill that we need to think about. But what
if we decided as a challenge to fill the glass using a completely "open-loop"
process? What would that look like?

A start might be to complete a detailed modeling effort, in which we nail
down the instantaneous water flow rate of water out of the pitcher, perhaps as
a function of the tilt angle. We would then need to know the detailed geometry
of of the glass. That is, for water flowing in a specified number of liters per
second the rate of rise of the level of water will depend strongly on the cross-
sectional area of the glass at the instantaneous water level. With all of this work
having been done very carefully, a true open-loop operation would mean that
we *close our eyes, tilt the pitcher at the predetermined sequence of angles,
with precise timings, and then stop*. If we have done our calculations correctly
and accounted for all possible effects (e.g., water lost to splashing if we are
particularly aggressive, minor variations in the proper gravitational constant
for our exact location on the earth), the end result will be a glass full according
to our agreed upon definition. Exactly.

One could imagine a conversation between a bemused nonengineer and the
triumphant engineer who oversaw solving this whole problem:

Spectator: What if someone gives you a different glass?

Engineer: No one is going to give us a different glass. The dimensions of this glass are compliant with an industry-wide standard for 500 mL glasses. To within a part per million, we're good. There's been talk of a "next-generation" 600 mL glass, but the last such upgrade took two years longer to roll out than everyone said. When it happens, we'll just spend a week or two redoing the model.

Spectator: Was it tough to automate the pouring process so that it was the same every time?

Engineer: After three months of practice, our original Pouring Lead still couldn't get his precision down to better than 5 percent error, and between that and the carpal tunnel syndrome we decided an industrial robot was the way to go. As a bonus, with its high-precision, high-speed pump we're no longer limited by gravity in terms of pour speed. This bad boy almost blew our budget, but it was worth it. As you are about to see.

Spectator: How fast can you fill the glass?

Engineer: 10.9 microseconds. That is the fastest fill time for a 500 mL glass ever reported in the literature and an industry first. A microsecond is what you get when you divide a second up into a million little pieces.

Spectator: But doesn't the water splash all over the place at that kind of pouring speed? In fact, doesn't the glass *explode*?

Engineer: Technically, yes. The impact of the water and the sonic boom that accompanies the pour are not kind to the glass. But if you integrate all of the water that lands within a 100 m blast radius of the pour site, we can show that we would have filled the the glass to within an error of only 1 percent. This is another industry record.

This example does illustrate one genuine advantage of open-loop systems over closed-loop systems: speed. For all of the over-the-top effort described in this short dialogue, it is true that when you fill a glass using your normal method of watching with your eyes until the glass is full, you are limited in your speed by your ability to react to what your eyes are telling you. So with closed-loop filling you can reliably fill any kind of glass with any kind of pitcher, here or on the moon, and with no modeling effort, *but you have given up speed*. This concept carries over to the feedback systems that are the subject of this book.

There are other concepts of closed-loop behavior that we can appreciate from looking at ordinary glass filling. At the start you pour quickly, for example, and then slow down as the water level gets closer to the final value (imagine the step response of a typical feedback system). Also the more accurate you want to be, the longer it will take you to complete the pour. (Note that settling time for feedback system gets longer as the settling tolerance gets more demanding.) Look for this behavior in closed-loop systems as the development continues.

2.2 Open- versus Closed-Loop Control in Block Diagrams

Jumping back to the language of linear systems, block diagrams for a closed- and open-loop system are shown in Figure 2.2. There will be a

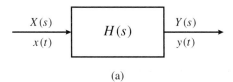

(a)

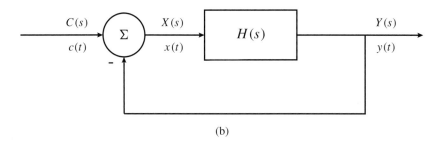

(b)

Figure 2.2 (a) Open-loop and (b) closed-loop systems.

lot of mathematics around these block diagrams, but a first, conceptual, and absolutely critical understanding can be gained right away. The central problem in control, remember, is how to craft the input of a system in order to get a desired output. That is, we may want our car to move at a certain speed, and $H(s)$ could be the transfer function between the force that our foot exerts on the accelerator and the speed of the car. Figure 2.2(a) shows the open-loop option. In order to get the desired output, we would begin by doing a thorough characterization of $H(s)$. Armed with this foreknowledge, we would create an $x(t)$ such that, when processed by $H(s)$, would result in the exact output that we want at exactly the time that we want it. Once we have determined the proper $x(t)$, the idea is that we shut our eyes and pass that $x(t)$ into the input of our system, trusting completely that we will get the desired output $y(t)$ based on our careful precharacterization of the system.

The feedback approach, illustrated in Figure 2.2(b), is very different. Consider the output of the summing junction, $x(t)$. It is an extremely interesting and valuable signal, because at any given time it is the difference between the the command, $c(t)$, and the desired output $y(t)$. Think about this: at every moment we know whether or not we have the desired output, and if not, how big our error is! If this is a cruise control in your car, for example, $c(t)$ might be the "command" of "go to 60 mph and stay there." If, when this command is issued, the actual speed of the car is 40 mph, then the signal $x(t_0)$ at that particular time is the (signed quantity) +20 mph, so the system "knows" that it is 20 mph too slow.

So we have the error at any given time. What do we do with it? Well, if your speed sensor was telling you that you were going 59 mph and you wanted to be going 60 mph, as a driver you would give the accelerator an extremely gentle nudge to move your speed up. On the other hand, if your speed sensor is telling you that your speed is 10 mph, you probably would not be so gentle. You would give a much more coarse, forceful push, and worry about honing your pressure on the accelerator more carefully when you were closer to your desired speed.

Very interesting: in this feedback type of approach, the response of you the driver varies in proportion to the size of the error. A crude model for this behavior is to follow the summing junction in Figure 2.2(b) with an amplifier whose output is exactly proportional to the error, as shown in Figure 2.3. Now suppose we know two things: 1) the system with our amplifier now in it is absolutely stable; 2) the DC gain of that amplifier is extremely high. What can we conclude? Consider a "DC" case, in which our car has been rolling at one speed for a long time and so we can use $H(0)$ in our analysis. If the final speed is y_0, we know that the final error e_0 is

$$e_0 = \frac{y_0}{H(0) \cdot A}. \tag{2.1}$$

That is to say, the larger our gain A, or the more "forcefully" we react to errors, the smaller our error will be.

An even more intriguing insight is that the larger our gain A, the closer $x(t)$ is to being *exactly the right input to $H(s)$ in order to get the correct output!* And we got this nearly perfect input not by doing a detailed model and relying on our ability to perfectly craft an input. We simply set up the system to automatically act to squash errors, and sat back and enjoyed the show.

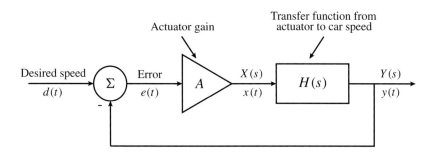

Figure 2.3 Simplified cruise control system for a car.

This, friends, is the power of feedback. And we're not done. There is more to appreciate if we look at a numerical example. Suppose that e_0 and y_0 are in miles per hour, and that $H(0) \cdot A$ is a dimensionless gain. Table 2.1 illustrates a miracle: where, with d_0 for the desired speed, we have used the algebraic result that

$$y_0 = \frac{H(0)A}{1 + H(0)A}d_0. \tag{2.2}$$

Table 2.1 *Ranges of k and resulting stability determination*

d_0	$|H(0) \cdot A|$	e_0	y_0
55 mph	10^6	0.000055	54.999945
55 mph	10^5	0.00055	54.999450
55 mph	10^4	0.0055	54.894500
55 mph	10^3	0.055	54.945000
55 mph	10^2	0.54	54.460000
55 mph	10^1	5.0	50.000000

This is astounding. We vary $H(0)$ by four orders of magnitude between 10^6 and 10^2, and our final speed varies only slightly, between 54.999945 mph and 54.46. And this is the true magic of feedback: the fundamental advantage of feedback is that it gives you extraordinary immunity to changes in the forward path. In the same way, using a form of feedback to pour water into a glass frees us from worrying ahead of time about the details of the glass.

Miracle? Yes. But is there a catch? Yes: we must take care of stability. We will spend much of the rest of the text developing this topic.

2.3 Anatomy of a Feedback Loop

In this section, we establish the conventions for talking about systems that we
will use for the rest of the book.

2.3.1 Block Diagrams

The block diagram is an extremely useful pictorial representation of linear
systems in general and feedback systems in particular. The rules governing a
particular block, depicted in Figure 2.4, are actually quite simple:

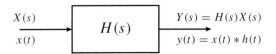

Figure 2.4 A basic block.

1. $X(s)$ is the input to the block. The block does not affect this input in any
   way. That is to say, $X(s)$ could be the input to one block or a million
   blocks at once; in the universe of block diagrams, it makes no difference.
   In the language of analog circuits, we would say that the block $X(s)$ has
   infinite input impedance. In the more general language of systems, we
   would say that there are no loading effects.
2. $Y(s)$ is the output of the block, as emphasized by the direction of the
   arrows in Figure 2.4. The output $Y(s)$ is $X(s)H(s)$. It bears emphasizing
   that $Y(s)$ could be fed to a million different blocks as their input, and yet
   according to the ground rules of block diagrams we need not account for
   this "load." In the language of analog circuits where voltages are the
   quantities of interest, we might say that the block $H(s)$ functions as an
   ideal voltage source with an output impedance of zero.

That's it, those are the rules. What is important to remember is just how
strong a simplification this abstraction represents. In reality, the boundaries
between parts or "blocks" of a system are never this neat and clean. It is
almost never true, for example, that you can take the output of one element
and drive the input of even one other block without altering the performance
of the driving block. The point is that *with intentional design*, this can be made
true to a good approximation. And so a block diagram remains a very useful
way to break a complicated system down into easily understood constituent
components.

There are a few elements that one encounters in block diagrams quite often and are worth taking special notice of. The first is a gain block, shown in Figure 2.5. A gain block is a special case of the $H(s)$ system block where

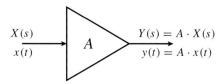

$$X(s)$$
$$x(t)$$

$$A$$

$$Y(s) = A \cdot X(s)$$
$$y(t) = A \cdot x(t)$$

Figure 2.5 Simple gain block.

the system function is simply the gain A for all frequencies. That means that in the frequency domain its response is absolutely flat in amplitude and phase "from DC to daylight," and in the time domain its impulse response is a scaled impulse. Its step response is an ideal step with an infinitely sharp rise time.

Another useful special case is the summing junction, shown in Figure 2.6. A summing junction does exactly what it sounds like, which is to produce at

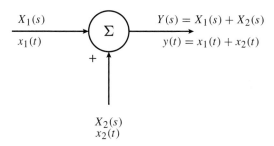

$$X_1(s)$$
$$x_1(t)$$

$$\Sigma$$

$$Y(s) = X_1(s) + X_2(s)$$
$$y(t) = x_1(t) + x_2(t)$$

$$+$$

$$X_2(s)$$
$$x_2(t)$$

Figure 2.6 A two-input summing junction.

its output the sum of its two or more inputs. The inputs and outputs have the same magical impedance properties that other, normal system blocks have. If one of the inputs is labeled with a minus sign, then that input is subtracted at the output instead of added. Note also that the frequency domain relation

$$Y(s) = X_1(s) + X_2(s) \qquad (2.3)$$

implies that in the time domain $y(t)$ is the instant-by-instant sum of $x_1(t)$ and $x_2(t)$. That is,

$$y(t) = x_1(t) + x_2(t). \qquad (2.4)$$

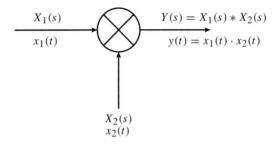

$$X_1(s)$$
$$x_1(t)$$

$$Y(s) = X_1(s) * X_2(s)$$
$$y(t) = x_1(t) \cdot x_2(t)$$

$$X_2(s)$$
$$x_2(t)$$

Figure 2.7 Multiplier block.

Finally, we have the multiplier, shown in Figure 2.7. This block is less common in systems, but you should expect to encounter it from time to time. When we invoke the multiplier in a system, what we mean is a block whose output at any instant in time is proportional to the arithmetic product of its two inputs at that same instant. Note that in the frequency domain the spectrum of the output is *not* the arithmetic, point-by-point product of the two input spectra. Rather, the output spectrum is the convolution $X_1(s) * X_2(s)$.

In an undergraduate course on feedback or linear system theory at this point, it would be typical to assign a lot of problems on block diagram manipulation. It's not a bad idea to work a few examples, and even one goes a long way. For example, study the block diagram in Figure 2.8.

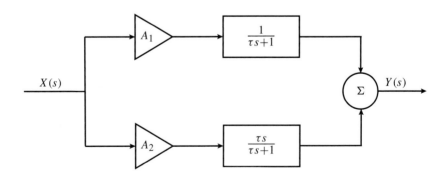

Figure 2.8 Example block diagram.

1. Write an expression for $Y(s)$ in terms of the input $X(s)$.
2. Approximate the transfer function $Y(s)/X(s)$ for $|s| \ll 1/\tau$. Repeat for $|s| \gg 1/\tau$.

3. Suppose that this block diagram represented an audio playback system, and that A_1 and A_2 were gain values set by the listener. We might call A_1 a "bass boost" and A_2 a "treble boost." Explain.

The readers are encouraged to work out this example, and find a few other examples to study from the recommended texts in the back of this book. But don't go overboard for now. You will get as fluent in block diagram manipulation as your chosen application demands.

2.3.2 Sensors and Actuators

It is easy in the study of feedback systems to get lost in the weeds of block diagrams and Laplace transforms, forgetting that these are representations of actual physical systems that we want to behave well. A brief look at sensors and actuators helps brings a healthy perspective.

One again, a car provides a useful example through its cruise control system. Let's look back at Figure 2.3 and get more into the details of the blocks. Figure 2.9 relabels the blocks to more clearly tie them to the cruise control function. Starting with the very left of the diagram, the "user input,"

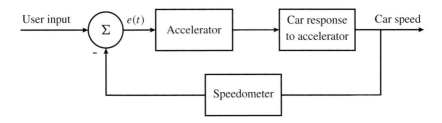

Figure 2.9 A cruise control block diagram, with slightly more detail.

we encounter the first significant issue in reducing these kinds of system ideas to practice. When switching on the cruise control, drivers have an idea in their head about how fast they want to go. At the time of this writing, the maximum allowable speed on US highways is 65 mph, so let us suppose that that is the desired speed. The first task of cruise control engineer is to give drivers a way of translating from the desired speed in their heads to a representation that is meaningful to an electromechanical system. One way to do this is to give drivers a means of representing the desired speed as a *voltage*. This might be done through a system of buttons on the driver's control panel, whose inputs to a microprocessor ultimate result in a voltage applied to the input of the speed control loop.

And what range of voltage might this input occupy? As we appear to have a lot of latitude in this choice, we can settle for now on a range between 0 V and 12 V, and modify that later if practical difficulties arise. Let us further imagine that federal safety regulations forbid the designers of cars of making cruise controls that permit speeds higher than 80 mph, and that it likewise does not make sense to have cruise control on for very low speeds, say, lower than 30 mph. Making these choices means that 0 V corresponds naturally to 30 mph, and 12 V corresponds to 80 mph. This means that the transfer function connecting the *desired speed* to the *input voltage to the control loop* is given as follows:

$$V_{\text{cmd}} = (S_D - 30)\, \frac{12\text{ V}}{50\text{ mph}}. \tag{2.5}$$

So input to our system diagram can be rendered in greater detail as Figure 2.10

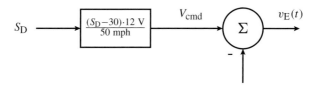

Figure 2.10 Translating desired speed to a control voltage.

It is worth pondering this extremely practical design step. Do not obsess over the implementation of the $(S_D - 30)$ 12 V/50 mph. Suffice it to say that is straightforward to set up an interface in which the user inputs a clear command in natural units, and internal digital-to-analog converter is deployed to generate the necessary voltage. The voltage summer is likewise straightforward.

It is after the summer where things get interesting, where we must work to translate physical reality to the sanitized world of block diagrams. This is a critical step, because we must capture the dynamics properly if we are to successfully assess the stability of the system that we design. The first step in the process is to properly model the kinematics of the car, which are determined by Newton's law $F = Ma$, or more expanded,

$$a_{\text{car}} = \frac{1}{M} \sum_n F. \tag{2.6}$$

That is, the acceleration of the car is equal to the sum of forces exerted on the car divided by M, the car's mass. And what forces are involved? The most obvious force involved is that provided by the motor by turning the wheels which, because of friction, result in a forward push on the car. We will call this F_m. We will suppose that this force is proportional to the degree to which we press the accelerator pedal,[1] and that in our automated control system we make this force proportional in turn to an internal voltage v_E. So the force that the motor ultimately exerts on the car, F_m, is for us

$$F_m = k \cdot v_E. \tag{2.7}$$

There are many other forces exerted on the car, such as friction in the internal moving parts and any slope (up or down) that the car happens to be on. To keep things simple, let's focus on one such force, wind resistance, and that we are moving slowly enough to experience viscous resistance:[2]

$$F_w = -\mu S_{car}. \tag{2.8}$$

The kinematic equation governing the speed of the car is thus given by

$$M a_{car} = k \cdot v_E - \mu S_{car}. \tag{2.9}$$

Now overall we are interested in the dynamics of a cruise control system, which governs the velocity of the car, not the acceleration. So it turns out to be helpful to cast relevant equations in terms of S_{car} instead of a_{car}. We can do so by remembering that

$$a_{car} = \frac{d}{dt} S_{car}. \tag{2.10}$$

The kinematic expression governing the speed of the car now becomes

$$M \frac{d}{dt} S_{car} = k \cdot v_E - \mu S_{car}, \tag{2.11}$$

and an important milestone in our development is complete. We have a differential equation that describes the important dynamics of the system, and we are now most of the way to a block diagram representation. The next step is to take the Laplace transform of both sides of Eq. 2.11, because the form of the resulting equation leads easily to a block diagram. We now have

$$M s S_{car}(s) = k \cdot V_E(s) - \mu S_{car}(s). \tag{2.12}$$

---

[1] Readers will note that we are doing an awful lot of "supposing," making educated guesses about how the system works. These guesses and assumptions should be checked, of course. But it is an immensely clarifying exercise to go first through this "rough draft" of the design to get one's mind around the critical, fundamental issues.

[2] Viscous resistance occurs when an object is moving slowly enough through a fluid (air in this case) to experience a retarding force proportional to the velocity. We don't have to go too fast before drag gets much worse, and goes as the *square* of the velocity.

We can solve for $S_{car}(s)$ to obtain

$$S_{car}(s) = \frac{k \cdot V_E(s)}{M(s + \mu)}. \tag{2.13}$$

We can at last write down the first part of our block diagram, the forward path, as shown in Figure 2.11. For now, the forward path contains the transfer function connecting the actuator input to the output variable that we care about. In this instance the actuator is the amalgamation of electromechanical parts that give rise to a torque exerted on the wheels by the engine, and the input is the voltage v_E. Note that we've done a little bit of looking ahead, in that

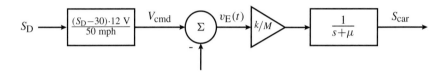

Figure 2.11 The forward path of our cruise control loop.

at this stage the summing block is not strictly necessary. But we know that we are going to be building a feedback system, and most feedback systems the actuator is driven with a signal that is linearly related to the instantaneous error (hence the "E" in v_E). We'll get that to now.

What is missing from this feedback loop is some kind of sensor, with which we will "close the loop." That is, we have our desired speed represented as command voltage v_{cmd}, and we now need some measurement of the actual speed we have achieved so that we can "know" if we have achieved our goal. Part of designing this sensor is making a decision about how it will represent its measurement internally to the system. Since we have been representing all variables as voltages so far, it is natural to imagine a speed sensor that represents speed in terms of voltage. Say that we buy this sensor, and that it is perfectly linear, and that it represents 0 mph as 0 V. Moreover, let's say that this sensor is capable of measuring speeds up to 120 mph, and that is represents this top speed with an output voltage of 12 V. The relationship between the measured speed and the output voltage is now

$$V_{meas} = \frac{12 \text{ V}}{120 \text{ mph}} \cdot S_{car} \tag{2.14}$$
$$= 0.1 \text{ V/mph} \cdot S_{car}.$$

Notice that right away, we have an issue with how we have set up our system. The problem is that we do not have agreement between our feedback and our command paths in terms of equivalence between speed in mph and voltage.

For example, in the command path 0 V corresponds to 30 mph, whereas in the feedback path 0 V corresponds to 0 mph. Also note that an increment of 1 mph in the command speed corresponds to a change of $0.1 \text{ mph} \cdot \frac{12 \text{ V}}{50 \text{ MPH}} = 24 \text{ mV}$, whereas in the feedback path a 1 mph change corresponds to a voltage increment of 100 mV. Pause to understand this as fundamentally a user interface problem: the feedback loop will always act to drive the error signal v_E to zero, and this lack of agreement means that our speed commands will be misinterpreted.

A reasonable first step is to adjust the command path to have the same incremental gain as the feedback path. Things like enforcing the limit that the command should never fall below 30 mph can be handled by a devoted, probably software-driven user interface. The new diagram is now shown in Figure 2.12.

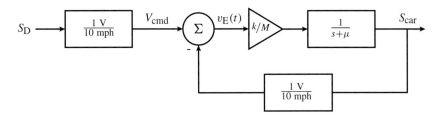

Figure 2.12 Next iteration of the cruise control block diagram.

After all of that, we finally have a block diagram that we can go back to our desks with and analyze. We are interested in the "dynamics" of the system: Is it stable? How fast does it take to settle to the commanded speed? Is its stability easily compromised by things like component tolerances? How accurate is it? The overall point of this exercise is to give the readers a start on being able to model sensors and actuators for the purpose of designing a feedback system. For this example the actuator was the electromechanical chain that reacts to an input voltage by exerting torque on the car's wheels, causing the car to accurate. The sensor takes a measurement of the car's speed, and represents that speed as a voltage. We note in passing that a phenomenal number of physical quantities can be probed by transducers whose response is detectable electronically. This is one reason that so many modern control systems are implemented electronically, as opposed to by mechanical engines.

We close this section by noting that it is extremely common, even normal, to get through a whole course on feedback and/or linear system theory and have not a clue about how to apply it to a real-world project. Don't be dismayed if this describes you. Studying this cruise control example is a good start.

2.3.3 Loop Transmission, Negative Feedback, and Stable Equilibria

We will find that something called the "loop transmission" is of absolutely central importance in the study of feedback systems. For now, consider what will become our generic feedback loop in Figure 2.13.

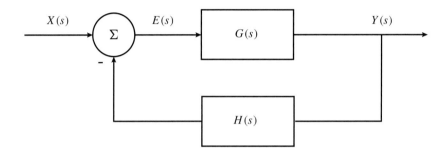

Figure 2.13 Our generic feedback loop block diagram.

The loop transmission is a transfer function that can be found by breaking the loop, injecting an input in on one side of the break, and determining the transfer function between that injected signal and the signal that comes back on the other side of the break. A picture helps here; see Figure 2.14. The loop transmission for this system is $Y_{\text{test}}/X_{\text{test}}$, which works out to be

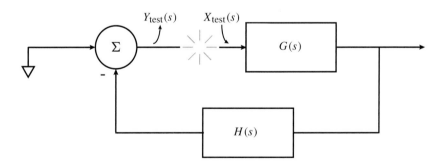

Figure 2.14 Breaking the loop and determining the loop transmission.

$G(s)H(s)$. Technically, of course, we have dropped a minus sign associated with the summing junction and the true loop transmission according to the given definition would be $-G(s)H(s)$. But dealing with such "negative feedback" systems is so common as to be the almost universal norm, and so the convention is to just identify $G(s)H(s)$ as the loop transmission without the minus sign.

The "negative feedback" nomenclature deserves some explanation. For simplicity, consider the case for which $H(s) = 1$ (e.g., unity for all frequencies), and a moment for which $x(t) > y(t)$. In the logic of our feedback control system, what we want is to drive our forward path in such a way as to increase y. With the minus sign where it is, the instantaneous error

$$e(t) = x(t) - y(t) \tag{2.15}$$

is greater than zero. So our actuator, represented by $G(s)$, will be driven by a positive signal. Moreover, the magnitude of that positive signal will be large for large errors, and small for small errors. If we assume that our actuator does not, roughly speaking, "have an inversion," then this positive error signal will drive the actuator in such a way as to increase $y(t)$.[3]

We thus arrive at the essence of "negative feedback." When perturbed from equilibrium, which is defined as *the error being zero*, a properly behaving negative feedback system responds in such way as to restore itself to this equilibrium, zero-error state. A good picture to keep in your mind is that of a ball at the bottom of a bowl, as shown in Figure 2.15.

By way of contrast, the essence of positive feedback is shown in Figure 2.16 and Figure 2.17. Figure 2.16 shows a feedback system where we have put in an integrator in the forward path. Notice that an equilibrium still exists. In the zero-error state, the integrator's output is still and therefore the system output does not change. However, in any number of scenarios $x(t)$ and/or $y(t)$ could change. In that instant $e(t)$ becomes nonzero, and the presence of the integrator combined with the lack of inversion at the summing junction ensures that the output $y(t)$ will go racing off to as close to infinity as the particular system will allow. Figure 2.17 illustrates this concept of an unstable equilibrium by way of a mechanical example.

So that minus sign, that inversion, is critically important. And its importance points the way to understanding the importance of the loop transmission. We spoke earlier in this section of the importance of an actuator doing nothing to

---

[3] This example is somewhat unfortunate, in that the longer you ponder it, the more nonsensical and problematic it gets. If your head starts to hurt from thinking about "signals running around the loop" *instantaneously*, and you feel that there must be a problem with this kind of analysis, then congratulate yourself on a deep level of insight. This is exactly the logical quandary that necessitates the entire formalism of feedback stability analysis. Rest assured, we are getting to that. In the meantime, mentally replace $G(s)$ with an integrator $\frac{k}{s}$ as a reasonable proxy for more general loop dynamics. With patience and continued careful thought, your headache will fade.

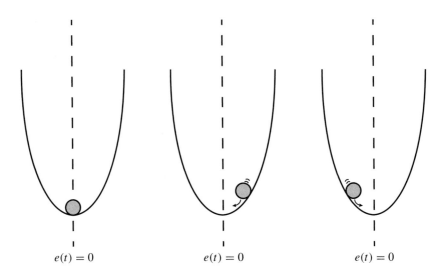

Figure 2.15 Illustration of a stable equilibrium. The bowl and gravity provide a natural restoring force that pushes the marble back to the equilibrium point.

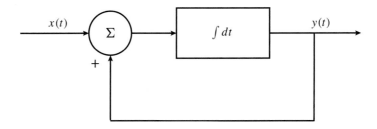

Figure 2.16 A system with positive feedback.

upset the inversion "roughly speaking." Well, what do we make of an actuator with the transfer function of Eq. 2.16?

$$G(s) = \frac{k}{(\tau_1 s + 1)(\tau_2 s + 1)(\tau_3 s + 1)}. \qquad (2.16)$$

If we look at very low frequencies, this actuator imparts almost no phase shift to an incident sinusoid. At one particular frequency, however, the actuator imparts a phase shift of exactly $-180°$ – which is an inversion! And for sinusoids at frequencies above that, the imparted phase shift is beyond $-180°$. What does that mean? What can we say about whether or not it is possible to close a stable feedback loop around this $G(s)$?

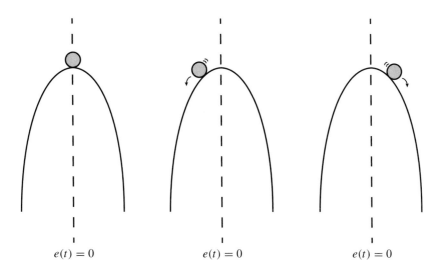

$$e(t) = 0 \qquad\qquad e(t) = 0 \qquad\qquad e(t) = 0$$

Figure 2.17 Illustration of an unstable equilibrium. Once perturbed, there is no
restoring force to return the marble to the equilibrium point.

This is the key question. For now, the point to understand is that because of
the importance of the inversion in the feedback loop, and because of the fact
that dynamics in the loop can cause inversions to appear and disappear on a
frequency-dependent basis, *of course* a detailed, frequency-domain characterization of the loop transmission is central to stability analysis. Pay attention
to this fact as we examine, in the coming sections, root locus techniques, the
Nyquist criterion, and derivatives of these techniques.

2.3.4 Black's Formula

Black's formula is an algebraic result that you will use so often that you should
commit it to memory. For the feedback system depicted in Figure 2.18, we have

$$E(s) = X(s) - H(s)Y(s) \tag{2.17}$$

and

$$Y(s) = E(s) - G(s). \tag{2.18}$$

Substituting for $E(s)$ in the second equation, we derive the transfer function
from the input $X(s)$ to the output $Y(s)$ as

$$\frac{Y(s)}{X(s)} = \frac{G(s)}{1 + G(s)H(s)}. \tag{2.19}$$

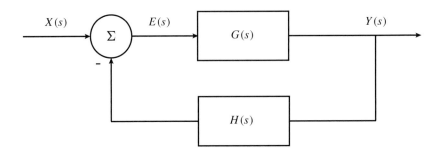

Figure 2.18 Illustration for deriving Black's formula.

This is the *closed-loop transfer function* of the system. Notice that the loop transmission $G(s)H(s)$ figures prominently in this expression, and we can rewrite the closed-loop transfer function as

$$\frac{Y(s)}{X(s)} = \frac{G(s)}{1 + L(s)}. \tag{2.20}$$

Here is another opportunity to appreciate the centrality of the loop transmission in the study of feedback systems. We notice that the poles of the closed-loop system are those values of s for which $1 + L(s) = 0$, and are therefore determined entirely by the loop transmission. A stable feedback system will have the roots of $1 + L(s)$ entirely in the left-half plane. An ideal oscillator will have roots of $1 + L(s)$ resting perfectly on the $j\omega$ axis. An unstable feedback system will have one or more roots of $1 + L(s)$ in the right-half plane. We will have more to say about right-half-plane poles in Section 2.4.3.

2.4 Delay Complicates Everything

We are slowly but surely working our way up to rigorous stability analysis techniques. The best way to appreciate these techniques is to first make some conceptual observations about what *must* matter for the stability of a feedback system. We started this process in Section 2.3.3, appealing to and strengthening your already sophisticated intuition about how a stable system should behave. We'll take the next step here by conceptually exploring delay in the loop transmission.

To imagine why delay must be important in feedback systems, consider a feedback process from your everyday experience: setting the water temperature in a shower. Imagine that you are in a hotel room, or at a relative's house, and are therefore unfamiliar with how long it takes for the water to warm up. You might guess a mixture of cold and hot water, put your hand under the tap, and feel that the water is very cold. What follows next for many people is that they crank up the hot water to the maximum, trying to get the water to heat up faster. After a delay of several seconds, the effect of the "hot water now!" command is fully and suddenly realized, and your hand shoots back protectively to avoid the overheated water. Fortunately by this point, there is no longer an appreciable delay between a faucet adjustment and a change in temperature of the water. *Absent this delay, you quickly and easily converge to a comfortable water temperature without any "overshoot."*

But now suppose instead that the delay never went away, and that there was always a 30-second delay between a faucet adjustment and the consequent water temperature change. What would your strategy be then? How would you deal with the delay in this feedback loop?

The answer is that you would converge to the final temperature very, very slowly. You would make very small adjustments, and you would wait for the full 30 seconds until you were sure that you knew the outcome of the adjustment before taking another step.[4]

Slowing down as a means of coping with delay is *fundamental* to feedback systems. As your understanding matures, make it a point to come back to things like the shower example and reconcile it with your growing analytical abilities. Have fun with them. How different would various competitions be if the average human reaction time was two seconds instead of 200 milliseconds?[5]

2.4.1 Phase Response as a Frequency-Dependent Delay

So we expect now that delay complicates feedback control. How does this show up mathematically? To start, if $x(t)$ is a time-domain signal, then $x(t-T)$ is that same signal delayed by T. To convince yourself of this, consider that the part of x that originally corresponded to $t = 0$ now appears at $t = T$. In general, the part that used to correspond at $t = t_0$ now corresponds to $t = t_0 + T$. Figure 2.19 Illustrates the concept.

---

[4] Actually, that's not quite true. Once you understood what was happening, you would lose patience and stop using feedback. You would get the water to the temperature that you liked once, and then mark the settings somehow so you could always return quickly to them. Open-loop control, as usual, is a much faster approach. Note that this strategy depends on the "DC gain" of the system being the *the same every time*. If you stuck with feedback, you would have some immunity to this variation in the gain of the forward path.

[5] Competitive chess aside.

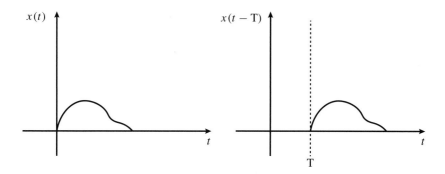

Figure 2.19 What a pure delay looks like for $x(t)$.

Next, we need the Laplace transform of $x(t - \mathrm{T})$. Start with

$$\int_{0_-}^{\infty} x(t - \mathrm{T})e^{-st}dt, \qquad (2.21)$$

and do the change of variables to $t' = t - \mathrm{T}$. This same integral can now be written

$$\int_{0_- - \mathrm{T}}^{\infty - \mathrm{T}} x(t')e^{-s(t'+\mathrm{T})}dt'. \qquad (2.22)$$

The limits of integration are at first inconvenient. But in real time control applications, it is often true (or it can often be made true for purposes of analysis) that $x(t') = 0$ for all $t' < 0_-$. Note also that $\infty - \mathrm{T}$ is, well, ∞. Rewriting now as

$$e^{-s\mathrm{T}} \int_{0_-}^{\infty} x(t')e^{-st'}dt', \qquad (2.23)$$

we have our big result, which is that the Laplace transform of $x(t - \mathrm{T})$ is very simply related to the Laplace transform of $x(t)$:

$$\int_{0_-}^{\infty} x(t - \mathrm{T})e^{-st}dt = X(s)e^{-s\mathrm{T}}. \qquad (2.24)$$

Similarly, for the Fourier transform we get $X(j\omega)e^{-j\omega\mathrm{T}}$.

So a delay leaves the Fourier transform *magnitude* unchanged. However, it does impart a phase to $X(j\omega)$ that linearly increases with frequency. The slope of that linear increase is the size of the delay, as seen in Figure 2.20. If you have ever studied feedback before, you know to get nervous when you see negative phase being piled on to the loop transmission. That said, with a singularity (a pole or a zero) you can't get this much negative phase! A pole in

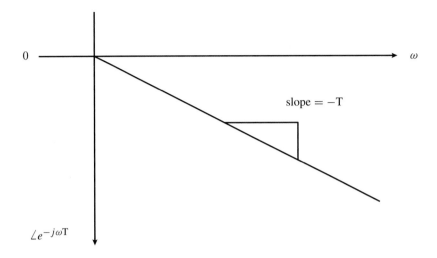

Figure 2.20 Phase as a function of frequency for a pure delay element.

the loop transmission can add at most $90°$, whereas you can see here that delay adds an unbounded amount of negative phase as the frequency gets higher.

This linear phase aspect of true delays make them a special case when it comes to phase responses. For a more general loop transmission transfer function $L(j\omega)$, we can express the result of passing sinusoid $e^{j\omega_0 t}$ through it as

$$|L(j\omega_0)|e^{j\omega_0 t}e^{-j\phi_L(\omega_0)},\qquad(2.25)$$

where the phase shift $\phi_L(\omega_0)$ is equal to $\angle L(j\omega_0)$. Rearranging the terms of Eq. 2.25 gives a highly suggestive new form:

$$|L(j\omega_0)|e^{j\omega_0 t}e^{-j\phi_L(\omega_0)} = |L(j\omega_0)|e^{j\omega_0\left(t-\frac{\phi_L(\omega_0)}{\omega_0}\right)}.\qquad(2.26)$$

When we compare this form to that of a general delay $x(t - T)$, we are reminded that the phase response of a transfer function is really a *frequency-dependent delay*. That is, the transfer function treats each sinusoid at its input as an individual, delaying each one according to its phase response.

Why bother to see phase response as a frequency-dependent delay? Well, if you are convinced by Section 2.4 that delay complicates everything, then you are well positioned to appreciate the phase response as a complete characterization of the delay through a transfer function. This is why the phase response figures so prominently in the root locus and Nyquist developments to come.

2.4.2 The Fundamental Oscillation Condition

An oscillator is a system that has poles exactly on the $j\omega$ axis. Such a system will have a ZIR of the form

$$y_{ZIR}(t) = C_1 e^{j\omega_1 t} + C_1^* e^{-j\omega_1 t} + C_2 e^{j\omega_2 t} + C_2^* e^{-j\omega_2 t} + \cdots \qquad (2.27)$$

For almost any set of initial conditions, the time evolution of such a system consists entirely of a superposition of nondecaying, nongrowing sinusoids. Usually when we think of oscillators we do so with systems that oscillate at a *single* frequency in mind, in which case we may say only C_1 (and C_1^*) in Eq. 2.27 are nonzero.

It is very common for feedback systems that are not properly designed to oscillate instead of carry out their intended control function. The *fundamental oscillation condition* is a condition on the loop transmission that, when satisfied, results in the closed-loop system exhibiting oscillations. Which is to say, when the fundamental oscillation condition is satisfied, the closed-loop system has poles on the $j\omega$ axis.

We refer once more to a generic closed-loop feedback system, now shown in Figure 2.21. The closed-loop transfer function is

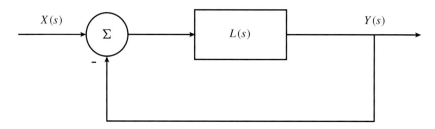

Figure 2.21 An even more generic feedback block diagram. Note that this form is equivalent in every way to Figure 2.18 if we have $L(s) = G(s)H(s)$ and place a block $1/H(s)$ between the input $X(s)$ and the input of the summing junction. We will often make this transformation going forward as we narrow our focus to the stability implications of $L(s)$.

$$H_c(s) = \frac{L(s)}{1 + L(s)}. \qquad (2.28)$$

If $H_c(s)$ has poles on the $j\omega$ axis, then it is necessarily true that for some ω_0,

$$1 + L(j\omega_0) = 0. \qquad (2.29)$$

This is a complete mathematical statement that captures what must be true of the loop transmission in order to support oscillation. It is customary to explicitly acknowledge the complex nature of $L(j\omega)$ and break Eq. 2.29 into two pieces, one governing its magnitude and one governing its phase. We say that a feedback system will oscillate at frequency ω_0 if the loop transmission satisfies

$$|L(j\omega_0)| = 1 \tag{2.30}$$

$$\angle L(j\omega_0) = -\pi.$$

The fundamental oscillation condition is extremely important. If an engineer remembers only one thing about feedback systems, this is it. But be careful. Most internalize it by thinking wrongly about sinusoids "running around the loop, reinforcing themselves." If this makes no sense to you, then good! Don't try to understand this line of reasoning, and spare yourself the trouble later of divorcing yourself from a false model that the human brain seems almost hard-wired to accept. If you find yourself nodding approvingly and saying to yourself, "That's exactly what I remember," then skip ahead to Section 4.2 before coming back here.

2.4.3 Poles in the Right-Half Plane Are Bad

A well-functioning cruise control system in a car is a fine example of a stable feedback system. The howling that results from a microphone getting too close to the speaker that it drives bespeaks an unstable feedback system. Very loosely speaking, stable feedback systems behave the way we want them to, faithfully following our commands until we exceed the bandwidth limit of the system. In contrast, unstable feedback systems go crazy as soon as you power them up.

We say that an LTI system is unstable if its transfer function has poles in the right-half plane. It is common to use a "pole-zero" diagram to represent poles and zeros pictorially. A pole-zero diagram is exactly what it sounds like: it has a real axis and an imaginary axis, and the poles and zeros of the transfer function in question are plotted there. For example, the transfer function

$$H(s) = \frac{(s + 10)^2}{s(s + 100)(s + 10 + 10j)(s + 10 - 10j)(s - 10)} \tag{2.31}$$

has the pole-zero plot shown in Figure 2.22. A common convention is that a pole is represented by a $\times$ and a zero by a $\bigcirc$. The right-half plane is exactly what it sounds like: it is the infinite plane that includes everything to the right of the $\sigma = 0$ line. This particular transfer function has a pole at $s = 10$, which is in the right-half plane (RHP). It follows that the system it represents is unstable.

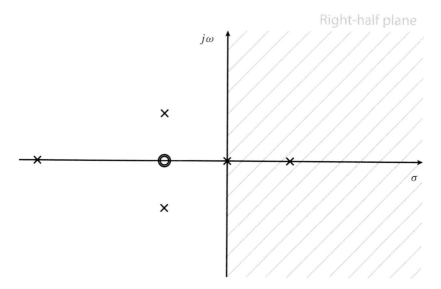

Figure 2.22 Pole-zero plot for Eq. 2.31. The right-half plane is the shaded region.

So how, exactly, does it "go crazy" upon power up? Well, the ZIR for this system will looking like

$$y_{ZIR}(t) = C_0 u(t) + C_1 e^{-100t} + C_2 e^{-10t} e^{j10t} + C_2^* e^{-10t} e^{-j10t} + \underline{C_3 e^{+10t}}.$$

$$(2.32)$$

It's that last term that is trouble, because it causes the ZIR to go rocketing off to infinity as time progresses, utterly overshadowing whatever else is going on in the system.

Right-half-plane poles occur in feedback systems when we fail to properly "compensate" them. Referring to Figure 2.23, if $G(s)$ represents the actuator and $H(s)$ the sensor and feedback path, $H_c(s)$ are the *dynamics that the designer introduces in order to get the desired behavior for the feedback system.* We call $H_c(s)$ the "loop compensation," or sometimes the "loop filter." Choosing the dynamics of $H_c(s)$ is called "compensating" the loop. Do we want the system to respond rapidly or slowly? Do we expect large manufacturing variation in $G(s)$, and therefore need the system to remain stable in the face of this variation? Is it important that the DC error of the control system be extremely small? These questions and more are what the designer answers by choosing the loop compensation $H_c(s)$. Root locus techniques and the Nyquist stability criterion are two tools that the designer has to ensure that whatever choice they make results in a stable feedback system.

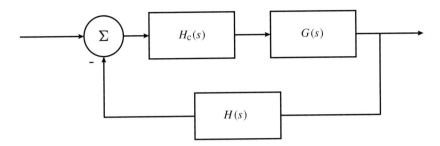

Figure 2.23 A feedback loop with a designer-supplied "compensator," or "loop filter."

2.5 Root Locus Techniques

The study of root locus techniques is a great introduction to the design of feedback systems. When undertaking such a design, or "compensating the feedback loop," you really have only two tools at your disposal. The first tool is introducing a (frequency-independent) gain in the system. The second tool is adding "dynamics" to the loop in the form of poles and/or zeros. The picture you should have in your mind is shown in Figure 2.24. Our convention will be

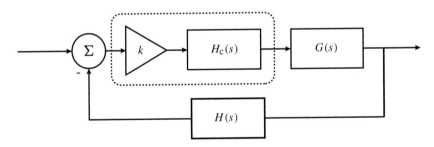

Figure 2.24 Separating the compensator into a "gain" part and a "dynamics" part.

that $H_c(s)$ has a DC gain of unity so that we cleanly separate the "gain" part of the compensation from the "dynamics" (the part that is frequency-dependent).

As we plunge into the design and analysis of feedback systems, it will be useful to use as few symbols as possible. We already know that the loop transmission is all that matters for stability, and going forward we will lump all the dynamics of these systems into $L(s)$. For example, referring to Figure 2.24:

$$L(s) = kH_c(s)G(s)H(s). \tag{2.33}$$

Whenever you are analyzing a feedback system, it is natural and useful to transform the original system into a generic, unity-gain feedback loop as shown in Figure 2.25. Take some time to convince yourself, using Black's formula,

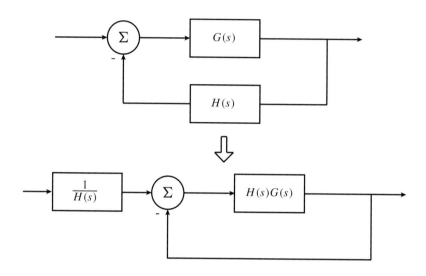

Figure 2.25 Redrawing a generic feedback loop as a unity-gain feedback loop.

that these systems are equivalent. Also note that the manipulation performed here, with the block $H(s)$ moving through the summing junction, can be validated on the basis of simple algebra. This is worth understanding and then *using*. The nice thing about transformation of Figure 2.25 is it provides a nice, visual separation between ideal closed-loop transfer function, $1/H(s)$, and the unity-gain feedback loop whose stability needs to be analyzed.

We'll actually make one more choice that makes analysis easier. It will be useful to keep a dimensionless gain k as a free parameter, and not lump it in with the rest of the loop transmission. So going forward, instead of as in Eq. 2.33 we will identify the loop transmission of Figure 2.24 as

$$L(s) = H_c(s)G(s)H(s) \qquad (2.34)$$

and focus our stability analysis on the modified block diagram of Figure 2.26.

Remember that the $1/H(s)$ block has not disappeared. When we go back to compute the closed-loop response to some stimulus, we must include it. For purposes of analyzing stability, however, all the essentials are contained in

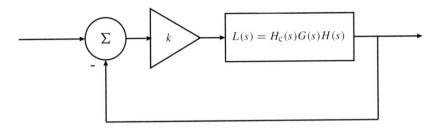

Figure 2.26 Redrawing Figure 2.24 as a unity-gain feedback loop for purposes of stability analysis. Note that we have dropped the $1/H(s)$ that properly belongs on the left side of the summing junction, as it is unnecessary for feedback stability analysis. However, we must put it back if we go to examine the complete closed-loop response.

Figure 2.26.[6] This will be our focus as we study root locus techniques and the Nyquist stability criterion.

2.5.1 The Problem We're Trying to Solve

It just so happens that an examination of Webster's definitions of "root" and "locus" brings some clarity to the study of root locus techniques. According to Webster, the relevant definition of a root is *a number that reduces an equation to an identity when it is substituted for one variable.* For purposes of this chapter, the equation that we care about is

$$1 + kL(s) = 0, \tag{2.35}$$

and the identity that it reduces to is $0 = 0$ when s is chosen to be a root. Roots of this equation are the all-important poles of the closed-loop feedback system.

Webster's definition of "locus" is *the set of all points whose location is determined by stated conditions.* In our case, the "stated condition" here is that $1 + kL(s) = 0$ for *some* value of k, and the "points" whose locations matter to us are points in the s-plane.

Putting these two definitions together, we arrive at a definition of "root locus." The root locus is simply *the set of all points in the s-plane that satisfy the equation $1 + kL(s) = 0$ for some value of k.*

A simple example goes a long way toward clarifying the problem that we are trying to solve. Consider the feedback system in Figure 2.27. The question

---

[6] We assume here that $H(s)$ is open-loop stable. Remember that the feedback path observes the output and, together with the summing junction, reports back how close that output is to the command input. If that feedback path itself is unstable, you have problems.

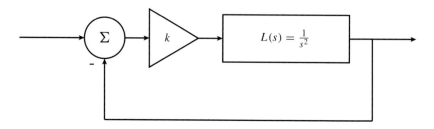

Figure 2.27 A first example for root locus analysis.

is, where are the closed-loop poles? Using Black's formula, we can write down the closed-loop transfer function easily as

$$\frac{Y(s)}{X(s)} = \frac{k/s^2}{1 + k/s^2} \tag{2.36}$$
$$= \frac{k}{s^2 + k}.$$

The closed-loop pole locations are where the denominator equals zero:

$$s = \pm j\sqrt{k}. \tag{2.37}$$

Evidently the pole locations depend on k. We can start to examine this dependence by constructing Table 2.2, a table of k values and the corresponding pole locations. In addition to representing it as a table, we can graphically represent

Table 2.2 A few values of k and the corresponding closed-loop pole locations

k	Pole locations
0	0, 0
1	$\pm j$
2	$\pm j\sqrt{2}$
10	$\pm j\sqrt{10}$

the dependence of the pole locations on k using a parameterized pole-zero plot, as shown in Figure 2.28. This figure shows that for any value of k, the closed-loop poles will remain on the $j\omega$ axis and therefore this system is an ideal

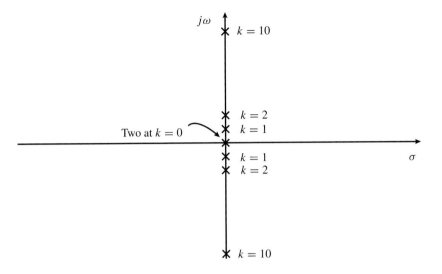

Figure 2.28 Precursor to a root locus plot.

oscillator. The frequency of oscillation is determined by the constant k: the larger k is, the greater the frequency of oscillation.

Now we can go one step further, which is to preempt the difficulties of maintaining ever more rows in our table or an ever more crowded pole-zero plot as we consider more values of k. What we do is graphically embrace the continuous nature of k, and sketch out in the complex plane the continuous *locus* of possible pole locations. Such a plot is called a *root locus plot*, and the example corresponding to our system is shown in Figure 2.29. The convention is that the arrows on the locus show the direction of migration for the closed-loop poles as k *increases*.

The promise of root locus techniques is that they will yield guidance and insight that is useful for design. What does this elementary root locus plot tell us about compensating this particular closed-loop system? We see quickly by looking at the root locus plot that if our goal is to build a stable control system we cannot do so if our compensation block is limited to a frequency-independent gain with no poles or zeros.[7] For *any* value of k that we choose, we will wind up with an oscillator whose oscillation frequency is determined by k. If we want to control this system, we will have to be more clever than that.

---

[7] So-called proportional control.

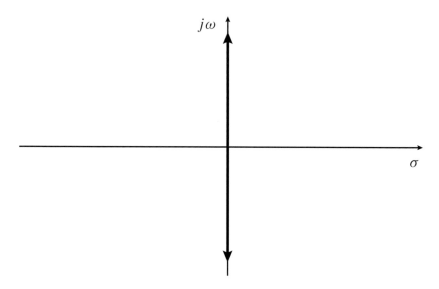

Figure 2.29 Our first, true root locus plot. The arrows on the loci show the direction in which the closed-loop poles move as k increases.

2.5.2 The Amazing Things You Can Do with Two Simple Conditions

There is a glaring hole in our otherwise promising development so far which is quickly exposed by a short experiment. Using only hand analysis, try to repeat the exercise of Section 2.5.1, but using the loop transmission

$$L(s) = \frac{(s+1)}{s^2(s+100)(s+200)(s+10^3)}. \tag{2.38}$$

Our new method doesn't look so promising now, does it?

Or at least it didn't, until the advent of powerful numerical solvers capable of factoring arbitrary polynomials in the blink of an eye. But back when root locus techniques were invented, such computational aids did not exist.

So big deal, right? Can't we just stop there, and rely on fast computers to generate root locus plots for us if we are careful? Why should anyone bother with the rest of this chapter, which helps you to sketch root locus plots quickly by hand?

Experienced engineers will smile ruefully at such questions. Computational tools are good and helpful, and do spare you from having to obtain exact numerical answers via your skill at a drafting table. But as any designer will tell you from painful, personal experience, there is *nothing* more hopeless than an engineer blindly surfing a simulator through a complex design space. If you are serious about designing things that work, you must have methods that give you conceptual insight into the design space. For feedback systems, root locus methods form an excellent toolset for deciding on compensation *strategy*. Can I add an integrator to $L(s)$ to drive down the DC error and still hope to stabilize the system? If k varies during the course of normal operation, will that result in closed-loop poles being pushed into the right-half plane? These are the types of questions that root locus methods help to answer quickly.

So the question is, how do we do a root locus plot with a complicated loop transmission, as in Eq. 2.38? First, let's make the difficulty explicit by writing out the closed-loop transfer function. Using Black's formula, we have

$$\frac{Y(s)}{X(s)} = \frac{k(s+1)}{s^2(s+100)(s+200)(s+10^3) + k(s+1)}. \qquad (2.39)$$

Even though we know the exact location of the open-loop poles, for the closed-loop poles we are left factoring a fifth-order polynomial that is parameterized in k. We have no choice but to throw ourselves onto a computer solver for answers, right?

Wrong. Computer solvers will always have an important place in engineering design. But it turns out that the locus of the closed-loop poles can be sketched by hand due to an astonishing number of implications of two simple conditions. In order for s_0 to be a pole of the closed loop system, it must satisfy

1. $|kL(s_0)| = 1$;
2. $\angle L(s_0) = -180°n$, where $n \in \{\ldots -3, -1, 1, 3 \ldots\}$.

Most root locus developments start this way, by defining two conditions, the magnitude and angle condition, in order. But it turns out that we can simplify things even more by observing that the angle condition is the key. If we can find a point s_0 in the complex plane for which the angle condition is satisfied, we then know that the angle condition is satisfied for a gain k given by

$$k = \left| \frac{1}{L(s_0)} \right|. \qquad (2.40)$$

So before diving into all of the rules for doing root locus plots, it will help you if you internalize the following definition of a root locus plot: *The root locus is the locus in the complex plane consisting of all points for which the angle condition is satisfied.* With just this rule, and a review of the geometric interpretation of poles and zeros (see Section 1.5.2), you can probably get pretty far on your own with root locus plots.

Take the example from Section 2.5.1, where we had a loop transmission of k/s^2. If we draw a pole-zero plot of the loop transmission, it looks like Figure 2.30. What we notice is that there are two regions in the s-plane where the angle condition can be satisfied: the $j\omega$ axis above the origin, and the $j\omega$

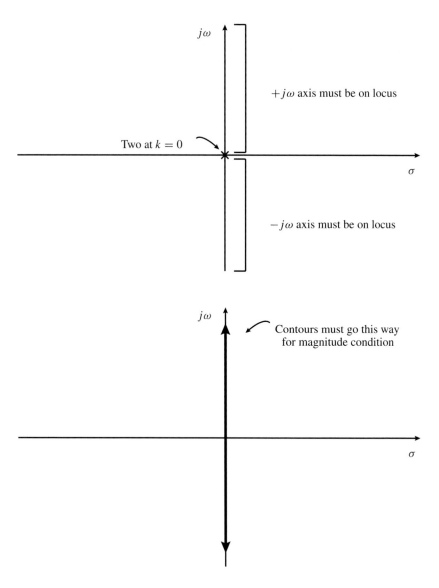

Figure 2.30 Arriving at Figure 2.29 using the angle and magnitude conditions.

below the origin. Why? Because the phase imparted by the poles of the loop transmission (located at the origin) is an odd multiple of $-180°$ only in these two places: $-180°$ along the $j\omega$ axis above the origin, and $-540°$ along the $j\omega$ axis below the origin. So right away, without having to construct any tables, we can immediately mark off the regions that belong on the root locus as shown in Figure 2.30. The only thing that we then must figure out is the direction the closed-loop poles migrate as the gain k is increased. Here, admittedly, the magnitude condition comes in handy. For small values of k, $|s|$ must also be small in order to satisfy $|k/s^2| = 1$. Similarly, for large values of k, $|s|$ must grow to keep $|k/s^2| = 1$. Evidently the arrows on the root locus plot point outward from the origin.

With the preceding example as a guide, an excellent exercise is to puzzle over the following examples. Get as far as you can, then compare with the answers at the end of this chapter. Most importantly, try to see clearly why your answer may differ from what is given. Finally, and this is critical, if you've got one wrong, try it again with the book closed and on a new, blank sheet of paper. This is an honest test of whether you've got it:

1. $L(s) = k/s$
2. $L(s) = k/s(s+1)$
3. $L(s) = k/(s+1)^3$
4. $L(s) = k(s+2)/s(s+1)$
5. $L(s) = ks/(s+1)^3$

A Short List of Root Locus Rules

We have come to the point where a few rules about plotting root locus plots are in order. It is actually not all that useful to start by memorizing these rules. In so doing, you run a considerable risk of missing out on the understanding that is so critical. It is far better to understand where the rules come from first, and then let the memorization come from their repeated application.

- **Rule 1** The number of branches, or paths of closed-loop poles, is equal to the number of open-loop poles.

Take the opportunity now to remind yourself of the thought experiment that is going on. We have a loop transmission of the general form $k \cdot n(s)/d(s)$, and the experiment is to vary the parameter k from near zero to infinity, mapping as we do so the locations of the closed-loop poles. These closed-loop poles are the roots of the characteristic equation

$$1 + k\frac{n(s)}{d(s)} = 0, \qquad (2.41)$$

which we can manipulate into the form

$$d(s) + kn(s) = 0. \tag{2.42}$$

Remember that the transfer function of a physical system has at least as many poles as zeros. This means the that the polynomial $d(s)$ is of order greater than or equal to that of $n(s)$. This in turn means that the order of $d(s)$ is equal to the order of Eq. 2.42, and we conclude that the number of roots of the characteristic equation is equal to the number of poles of the open-loop system. The simple fact of closing a feedback loop around a system does *not* alter the number of poles.

There is something very satisfying about this humble rule. While feedback does not add or subtract poles to a system, the dynamics of the closed-loop system are otherwise profoundly altered. It is often true, for example, that the closed-loop system will settle much faster in response to a step than the open-loop system would. Lest you think that you have gotten something for free,[8] what we have usually given up is gain. That is a trade-off that is not really visible in root locus plots, but look out for this moving forward.

- **Rule 2** The branches of a root locus plot begin, for values of k near zero, at the open-loop pole locations, and end, as k approaches infinity, at the locations of the open-loop zeros.

This rule actually gets a little confusing unless you remember that transfer functions with more poles than zeros are considered to have zeros at infinity. That is, $1/s$ has a pole at the origin and one zero at infinity; $(s + 1)/(s + 2)(s + 10)$ has two poles, one zero at -1, and one zero at infinity.[9]

That said, validation of Rule 2 is found through considering the magnitude condition. We require

$$|kL(s)| = 1. \tag{2.43}$$

For k near zero, $|L(s)|$ must be correspondingly close to infinite to satisfy the magnitude condition. It follows that for small k, the locus in in the vicinity of the poles of $L(s)$. Similarly, for k approaching infinity, $|L(s)|$ must be correspondingly close to zero. It follows that the branches of the root locus terminate on the zeros of $L(s)$.

---

[8] In engineering, we *never* get something for free.
[9] In the limit of large s, $(s + 1)/(s + 2)(s + 10) \approx 1/s$, and so "has a zero of multiplicity one at infinity." Similarly, the transfer function $(s + 1)/(s + 2)(s + 10)(s + 100)$, which is $\approx 1/s^2$ in the limit of large s, is said to have two zeros at infinity, or "a zero of multiplicity two at infinity."

- **Rule 3** Branches of the root locus lie on the real axis to the left of an odd number of real poles and zeros.

First, remind yourself that complex conjugate pole pairs contribute nothing to the phase of points along the real axis. From there, realize that Rule 3 is a simple consequence of the angle condition:

$$\angle L(s_0) = -180° \cdot n, n \in \{\ldots -3, -1, 1, 3\ldots\}. \qquad (2.44)$$

Figure 2.31 illustrates Rule 3.

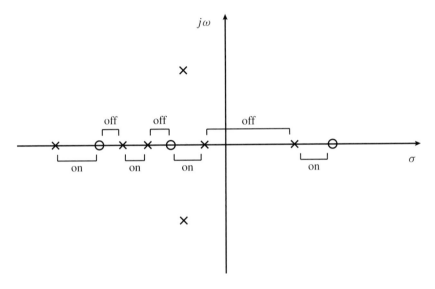

Figure 2.31 Illustrating Rule 3 for root locus plots: branches of the root locus lie on the root axis to the left of an odd number of real poles and zeros. Note that the complex pole pair in this figure is irrelevant for the application of Rule 3.

- **Rule 4** If a branch lies on the real axis between two poles, the locus must break away from real axis somewhere between those two poles. Similarly, if a branch of the locus lies between two zeros, there must be an entry point between the zeros.

Figure 2.32 shows an examples of Rule 4. In each case, convince yourself using the geometric view of poles and zeros that the drawn locus is reasonable.
   One more:

- **Rule 5** As k gets large, $P - Z$ branches of the locus go off to infinity (Rule 2), where P is the number of open-loop poles and Z is the number of

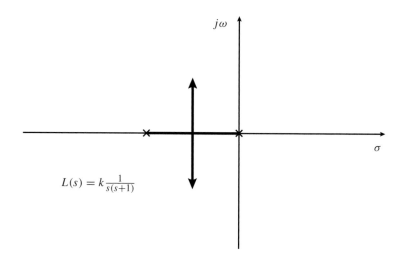

$$L(s) = k\frac{1}{s(s+1)}$$

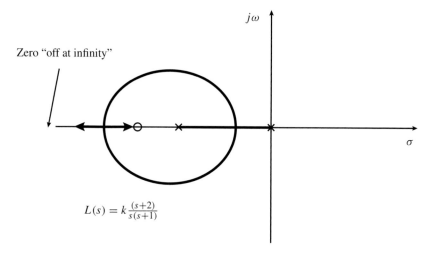

Zero "off at infinity"

$$L(s) = k\frac{(s+2)}{s(s+1)}$$

Figure 2.32 Illustrating Rule 4 for root locus plots. For the bottom figure, we say that $L(s)$ has two real poles, one real zero, and one zero "at infinity," because in the limit of large $|s|$ the transfer function looks like $\frac{k}{s}$. By the same logic, the loop transmission $\frac{1}{s(s+1)}$ is said to have two zeros at infinity. You might be justly concerned that these transfer functions go to zero for large $|s|$ regardless of the *angle* of s. For purposes of Rule 4, though, use Rule 3 to keep yourself grounded.

open-loop zeros. These branches that go off to infinity approach asymptotes
that have angles α_i to the real axis given by

$$\alpha_n = \frac{(2n + 1) \cdot 180°}{P - Z}, \qquad (2.45)$$

where $n = 0, 1, \ldots, (P - Z - 1)$, and the centroid σ of the asymptotes lies
on the real axis at

$$\sigma = \frac{\Sigma p_i - \Sigma z_i}{P - Z}. \qquad (2.46)$$

That's the short list of root locus rules, and enough to get started. For the
curious, a more complete list of root locus rules can be found in any one of the
feedback texts listed in Chapter 6. It bears repeating that the angle condition is
a reliably excellent guide to finding one's way through root locus plots. If you
remember only that first, and the magnitude condition second, then the major
features of even the most complicated root locus plots will be understandable.

2.5.3 Root Locus as a Design Tool

The hard thing about root locus is that, from the standpoint of hand calculations, it is very difficult to get numerical results. Imagine doing a root locus
plot, and discovering to your surprise and delight that if the gain k is tuned just
right, you can put your closed-loop poles exactly where you want them to be.
You discovered this, again in your imagination, by fluently applying all of the
rules of root locus plot drawing. Now that you know that a suitable value of k
exists, what is the numerical value of k?

Exactly. You have no idea, and there's nothing in the root locus methodology that has equipped you to even begin to answer that question.

The other reason root locus plots are difficult is that it is rarely practical
to draw them to scale. It is not uncommon, for example, for an open-loop
system to have poles that are several decades apart (a pole in the tens of
Hz, for example, and then poles scattered from 100 kHz to 1 MHz and
beyond). Capturing such a state of affairs on a pair of linear axes is completely
impractical. The right attitude to take about root locus plots is that they are
cartoons, even caricatures, that highlight the most outstanding features of a
given system in a way that is useful for design.

A good root locus plot can be a tool that tells us whether a design concept
for stabilizing a feedback system is sound or not. For example, suppose that
the element that we are trying to control through feedback is well modeled
by a single, low-frequency pole and a DC gain of a_0, shown in Figure 2.33.

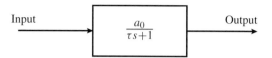

Figure 2.33 Simple model of a system block that we will control using a feedback loop.

If we are "controlling this block using feedback," what we mean is that we will include this block in the forward path of a feedback loop according to Figure 2.34, and choose $H(s)$ such that the system output follows the command input according to our dynamic criteria.

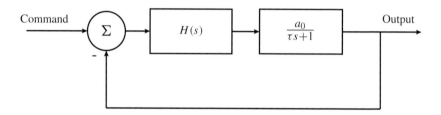

Figure 2.34 Closing the loop around the block of Figure 2.33.

Suppose that what matters to us is to have high speed following, as high as possible, while still ensuring that there is no DC error in response to a step. As we will be reminded in Section 2.6.2, we can get zero DC error by placing a pole at the origin in the forward path. As a first step in our exploration of the design space, then, what we might do is put a pole at the origin, and see what happens on the root locus plot.

The root locus plot is shown in Figure 2.35. (Note the application of rules 3 and 4 from our short list.) First, the good news: for any choice of k, we never get a pole in the RHP! This means, technically, that the system is always stable. We also know that by placing a pole at the origin, we have won the battle against DC errors. But what about other aspects of the dynamic response? Are there any sensible limits on the value of k?

It turns out that this is a good time to get acquainted with how a complex pole pair in the closed-loop response causes a feedback system to behave. In this particular example it is clear that the closed-loop behavior will be entirely determined by a single complex pole pair, but it will often be true in real life that the behavior of a feedback system is well approximated by that of a single complex pole pair. Let's explore this.

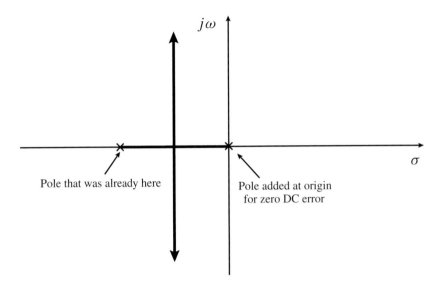

Figure 2.35 Using an integrator as the compensator and closing a loop around the block of Figure 2.33.

Why Are So Many Feedback Systems "Dominated" by a Single Complex Pole Pair?

We have talked of root locus techniques as a way of evaluating strategies for designing feedback systems. What we are missing is a basis for evaluation of a root locus plot: does the strategy in question result in a stable system with "acceptable" characteristics?

It will turn out that the behavior of many feedback systems is dominated by either a single closed-loop pole, or a single complex pair of closed-loop poles. The acceptability of the system behavior will in turn by determined by the geometry of these dominant poles on the pole-zero diagram. We must therefore address the question of "dominance" of a pole or pole pair, and then the characteristics that these poles imbue to the behavior of the closed-loop system.

First, the question of dominance. Consider a transfer function with two left-half-plane poles, where we have normalized the transfer function such that its DC gain is unity:

$$H(s) = \frac{p_1 p_2}{(s + p_1)(s + p_2)} \qquad (2.47)$$

$$= \frac{p_1}{s + p_1} \cdot \frac{p_2}{s + p_2}.$$

Figure 2.36 A two-pole system.

A block diagram of such a system is shown in Figure 2.36. Now, consider the step response of such a system. If $h(t)$ is the impulse response of the system, it can be shown that the step response $h(t) * u(t)$ is

$$h(t) * u(t) = \frac{p_1 p_2}{p_2 - p_1} \left(\frac{1}{p_1}(1 - e^{-p_1 t}) - \frac{1}{p_2}(1 - e^{-p_2 t}) \right) u(t). \quad (2.48)$$

The key from this point on is to realize that in many common systems, it just so happens that the poles tend to be widely spaced, or there tends to be one or two poles that are much slower than all of the rest. If that is the case, say, $|p_1| \gg |p_2|$, then Eq. 2.48 simplifies to

$$h(t) * u(t) \approx (1 - e^{-p_2 t} u(t)). \quad (2.49)$$

That is, the time-domain step response looks very much as though it were that of a single-pole system, where that pole corresponds to the *slower* of the two poles. Checking: for $|p_2| \gg |p_1|$,

$$h(t) * u(t) \approx (1 - e^{-p_1 t} u(t)). \quad (2.50)$$

Again, the *slower* pole "dominates" the dynamics of the system. Over and over again, you will find that the poles and zeros closest to the origin are the most "important" in a system.

For the inexperienced, there may at first be some frustration when seeking to apply this insight. One question: what degree of inequality satisfies $|p_1| \gg |p_2|$? The answer is that this is the approximator's art and it depends on how accurate you want to be in your analysis. A useful starting point is that you can assume that pole i dominates pole j if $|p_i| < 1/10 \times |p_j|$. That is, if the poles are separated by an order of magnitude, the slower one is the one that matters most.

Now that we know what it means for a pole or pole pair to "dominate," let's return to the question of why it is so common for closed-loop feedback systems, for all their apparent complexity, to appear well approximated by a single pole or complex pole pair. Start by understanding that that if you are even bothering with the details of feedback theory, there is an inherent aggressiveness in your design approach. You are saying, in effect, "How can

I absolutely maximize performance of this feedback system, subject to the constraint of keeping it stable?" When you stop optimizing, you are almost always doing so because you have reached a point where a closed-loop pole or pair of poles are as close to the right-half plane as you can tolerate. It is normally true that these closed-loop poles, in their proximity to the right-half plane, are *also much closer to the origin than all of the other singularities.* It is because they are closer to the origin that they dominate the closed-loop behavior.

Stability and the Characteristics of Second-Order Systems

For reasons described in the previous section, it will often happen that complicated closed-loop systems behave very much as though they are second-order. That is, the closed-loop transfer function is well approximated by

$$H(s) = \frac{1}{(s + \sigma_0 + j\omega_0)(s + \sigma_0 - j\omega_0)}. \tag{2.51}$$

It is common to normalize transfer functions so that their DC gain is unity. Doing that here, the result is

$$H(s) = \frac{1}{\frac{1}{\sigma_0^2 + \omega_0^2} s^2 + \frac{2\sigma_0}{\sigma_0^2 + \omega_0^2} s + 1}. \tag{2.52}$$

This is mathematically correct, but it is an ungainly and awkward description of a system that will show up over and over again in the analysis of linear systems. What we seek is a shorthand, a way of looking at a system dominated by complex pole pair and describing its behavior without the need to resort to a complicated equation. The purpose of this small section is to derive commonly used parameters used to describe second-order systems, and relate them to the transfer function.

If you kick a second order system with an impulse, the system will respond according to

$$h(t) = e^{-\sigma_0 t} \sin \omega_0 t. \tag{2.53}$$

It turns out that the way that this system behaves is really determined by the size of $|\sigma_0|$ relative to $|\omega_0|$ (assume that $\sigma_0 > 0$). Consider what happens to the exponential over the time scale determined by the period of the oscillatory term. For the period, we have

$$T = \frac{2\pi}{\omega_0}. \tag{2.54}$$

We are interested specifically in how much the amplitude of the oscillation dies down relative to the period of that oscillation. The amplitude $A(t)$ is $e^{-\sigma_0 t}$, and after one period T the amplitude is given by

$$A(T) = e^{-2\pi\left(\frac{\sigma_0}{\omega_0}\right)}. \tag{2.55}$$

Now it is easy to see the importance of the relative magnitudes of σ_0 and ω_0. If $|\omega_0| \gg |\sigma_0|$, the impulse response will look like Figure 2.37. We call

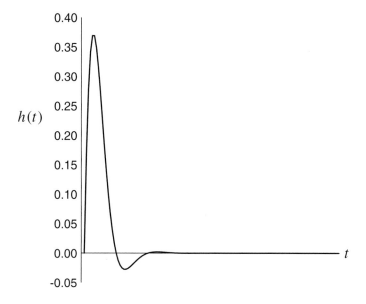

Figure 2.37 Impulse response (not step response) of a heavily damped second-order system.

this type of response "heavily damped," because there is very little "ringing." That is, the oscillatory response is overwhelmed by aggressive, exponential decay of the envelope. In contrast, if $|\omega_0| \ll |\sigma_0|$, then after one period T the amplitude $A(t)$ has hardly died down at all and what we see is the response in Figure 2.38. We call this response "lightly damped" or exhibiting "a lot of ringing." An extreme example of a lightly damped system is a tuning fork, whose period of oscillation might be on the order of millisecond but which might take 30 seconds or more after being struck to become inaudible.

Once at peace with the importance of the relative magnitudes of σ_0 and ω_0, we can return to a pole-zero diagram and gain fresh insight. Look at the three cases shown in Figure 2.39. What jumps out right away is that small values of θ

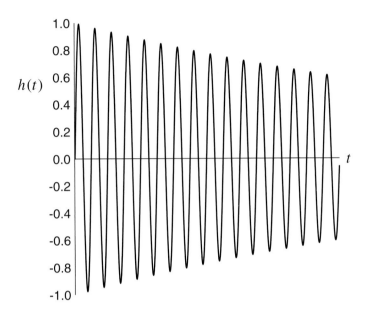

Figure 2.38 Impulse response (not step response) of a lightly damped second-order system.

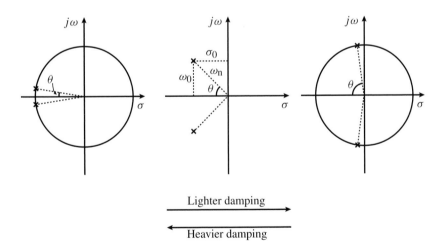

Figure 2.39 The geometry of a complex pole pair and the damping ratio. At just the point where $\theta = 0$ and the two poles coincide on real axis, the system is said to be "critically damped."

correspond to more heavily damped systems, whereas values of θ approaching 90° correspond to lightly damped systems.

Armed with this insight, we return to Eq. 2.52 with the goal of making the form even a little bit less obscure. At first, θ appears nowhere to be found. But after dusting off our old trigonometry textbooks and referring to Figure 2.39, we notice that $\sigma_0^2 + \omega_0^2$ is the square of the hypotenuse of a right triangle whose sides are of length $|\sigma_0|$ and $|\omega_0|$. We follow the convention of the literature of calling $\sqrt{\sigma_0^2 + \omega_0^2}$ the *natural frequency*, or ω_n.[10]

Continuing to lean on our trigonometric identities, we note that $\cos\theta$ can be expressed as

$$\cos\theta = \frac{\sigma_0}{\sqrt{\sigma_0^2 + \omega_0^2}} = \frac{\sigma_0}{\omega_n}. \qquad (2.56)$$

We are almost to the form of Eq. 2.52 that appears in virtually all textbooks on linear systems. Having defined $\cos\theta$, Eq. 2.52 becomes

$$H(s) = \frac{1}{\frac{s^2}{\omega_n^2} + \frac{2}{\omega_n}\cos\theta \cdot s + 1}. \qquad (2.57)$$

The final step is to assign $\cos\theta$ to the Greek symbol ζ, which we call the "damping ratio." We have at last the canonical form for a second-order transfer function:

$$H(s) = \frac{1}{\frac{s^2}{\omega_n^2} + 2\frac{\zeta}{\omega_n}s + 1}. \qquad (2.58)$$

Going forward, remember when ζ ranges between 0 and 1: $\zeta = 1$ corresponds "critical damping" with two poles coincident on the real axis; $\zeta = 0$ corresponds to no damping, with two poles on the imaginary axis. When $\zeta > 1$, the situation corresponds to two poles on the real axis, and there is no oscillatory part of the response.

There are a number of handy relationships that relate characteristics of the step response Figure 2.40[11] and frequency response Figure 2.41 to the damping ratio ζ and natural frequency ω_n.

Further explanation can be found in Thomas Lee's excellent *The Design of CMOS Radio-Frequency Integrated Circuits*.[12] Very briefly, here is a list of useful relationships:

---

[10] Why "natural frequency"? Well, the units are right …
[11] Note that this figure deals with the *step* response, whereas Figures 2.37 and 2.38 deal with the *impulse* response.
[12] Thomas Lee, *The Design of CMOS Radio-Frequency Integrated Circuits* (Cambridge, UK: Cambridge University Press, 1993).

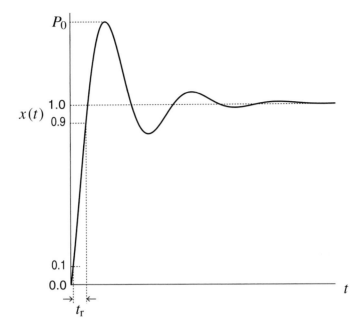

Figure 2.40 Second-order step response. This figure is the reference for Eqs. 2.59–2.61.

$$t_r \approx \frac{2.2}{\omega_h}. \tag{2.59}$$

Equation 2.59 relates the rise time (t_r) of the transient step response to the "3-dB bandwidth" of the system. You would expect that the higher the bandwidth, the shorter the rise time. By convention, the rise time is measured from the time the response gets to 10 percent of its final value until the time it first reaches 90 percent of its final value:

$$P_0 = 1 + \exp\left(\frac{-\pi\zeta}{\sqrt{1-\zeta^2}}\right). \tag{2.60}$$

Equation 2.60 expresses the percentage overshoot in the step response in terms of the damping ratio. As ζ approaches unity, note that there is no overshoot and therefore P_0 approaches 1:

$$t_{s2\%} = \frac{4}{\zeta\omega_n}. \tag{2.61}$$

Equation 2.61 is the "settling time," or $t_{s2\%}$, is the time it takes for the step response to settle to within 2 percent of its final value:

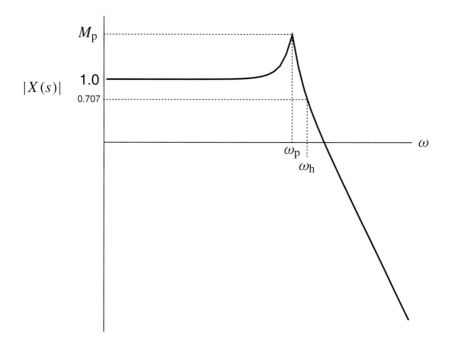

Figure 2.41 Second-order frequency response. This figure is the reference for Eqs. 2.62–2.64.

$$M_p = \frac{1}{2\zeta\sqrt{1-\zeta^2}} \quad \left(\zeta < \frac{1}{\sqrt{2}}\right). \tag{2.62}$$

The M_p of Eq. 2.62 is called *magnitude peaking* in the frequency response. It is well characterized by this expressions for light to moderate damping, but becomes invalid when ζ exceeds about $1/\sqrt{2}$:

$$\omega_p = \omega_n\sqrt{1-2\zeta^2}. \tag{2.63}$$

In Eq. 2.63, ω_p is the frequency at which the magnitude peak occurs:

$$\omega_h = \omega_n\left[1 - 2\zeta^2 + \sqrt{2 - 4\zeta^2 + 4\zeta^4}\right]^{\frac{1}{2}}. \tag{2.64}$$

Equation 2.64 relates the 3-dB bandwidth, ω_h, of the system to the natural frequency and the damping ratio. As with all of these expressions, try out the limits of the damping ratio (at zero, and then at 1) just to see if they behave the way that you expect.

Also, take a moment to go back to the example of Figure 2.35. We noted at the time that we could increase k seemingly without limit, and still technically have a stable system. We asked, in fact, if there were any sensible limits on the value of k. With your knowledge of second–order systems refreshed, you can go back now and see that two major things happen as k increases: first, the system gets faster as ω_h increases; second, the complex pole pair has a smaller and smaller damping ratio ζ. A small damping ratio means lots of ringing in the step response, and lots of peaking in the frequency response, and both are usually undesirable in a control system. Going forward, we will equivalently describe systems whose dominant complex pole pair is lightly damped as having "small stability margins." Think about this.

We close this section on the behavior of second-order systems with a reminder to not lose sight of the forest because of the trees. The "trees" in this case are all of the math surrounding second-order equations. The "forest" is the following thought sequence: we often observe in physical systems that their behavior in response to stimulus appears to be fully determined by a single, complex pole pair; curious, we looked analytically at why this is the case; convinced by both experiment and analysis that second-order systems are everywhere, we developed a convenient shorthand around concepts like the damping ratio, natural frequency, percentage overshoot, and frequency peaking. This shorthand is simply convenient in the everyday work of discussing second-order systems. For example, knowing all of the foregoing we can now look at a root locus plot and estimate behavior based on the geometry of the dominant closed-loop pole pair.

2.5.4 Root Locus in Discrete Time

Root locus analysis is fun and interesting, but in real life it is sometimes complicated by the fact that we quite often do not have the complete pole-zero map of our system components at hand. Op-amps and motors, for example, almost never have their poles and zeros specified as part of their data sheets. So in real life, before you can even do root locus analysis, there is usually a detailed modeling step wherein you must sniff out approximations to the relevant system functions.

An exception to this rule occurs when all aspects of the system are neatly contained in the internals of a discrete-time signal processor. We need only look at the code and/or the schematics to know *exactly* where all of the poles and zeros lie. Root locus techniques shine here, and so it is worth spending a moment on how to apply root locus in this context.

It turns out there are two aspects to the discussion. The first concerns how the closed-loop poles move in response to varying the parametric gain k, and the second concerns interpreting the resulting pole-zero diagram.

Root Locus Rules

Figure 2.42 shows an example diagram for a discrete-time feedback control system. The open-loop poles are $\alpha = r_\alpha e^{j\Omega_\alpha}$ and $\beta = r_\beta e^{\Omega_\beta}$. The open-loop transfer function is

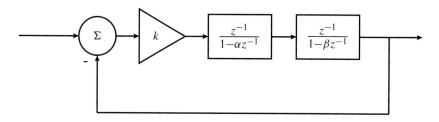

Figure 2.42 A DT feedback system.

$$\frac{kz^{-2}}{1 - (\alpha + \beta)z^{-1} + \alpha\beta z^{-2}},\qquad(2.65)$$

which we can write as

$$\frac{k}{z^2 - (\alpha + \beta)z + \alpha\beta}.\qquad(2.66)$$

If we ask ourselves how the closed-loop poles behave as k varies, and if poles are stil the values of z for which the closed-loop transfer transfer function denominator vanishes, then ... the problem is *exactly* the same as it was when we first took up root locus in continuous time. *All* of the same rules apply! The root locus rules from continuous time can be applied to discrete time with no change. For example, if $\alpha = 1/4$ and $\beta = 1/2$ in this example, then the root locus plot will look like Figure 2.43. What differs is how we interpret this plot. In continuous time, with the poles already in the RHP, we would say that this system is unstable. How does it work in DT?

Interpreting Root Locus Plots in DT

In CT, the imaginary axis divided the complex plane into two regions. To the left of the imaginary axis, we had complex exponentials that, oscillate how they will, were doomed to die off exponentially in time. The right-half plane defined exponentials that grow without bound. The imaginary axis itself defined a

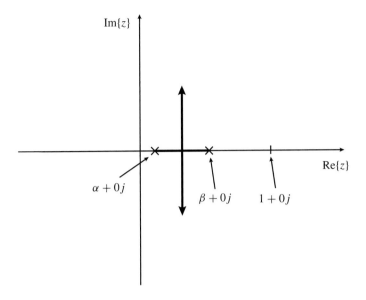

Figure 2.43 A root locus plot for the example DT system.

strange, complex exponential purgatory, in which exponentials neither grew nor died. They just stayed wiggling around nervously ...

Things are different in DT, but only because the mathematical form of the basic complex exponential is different. In DT, a complex exponential takes the form

$$x[n] = r_0^n e^{j\Omega_0 n}. \tag{2.67}$$

In this form, we can determine what exponentials will grow geometrically with no reference to the complex part. Very simply, if $|r| > 1$, we have a problem. If $|r| < 1$, the exponential will die as n increases. And if $|r| = 1$ exactly, we have

$$x[n] = \cos \Omega_0 n + j \sin \Omega_0 n. \tag{2.68}$$

The result is that for DT, a circle of unit radius is what separates growing exponentials from dying ones. Figure 2.44 illustrates this concept.

If we go back to the example from Section 2.5.4, we can see immediately how the root locus plays out in Figure 2.45. Right away, we see that the system starts out as stable for for small values of k, and goes unstable for sufficiently

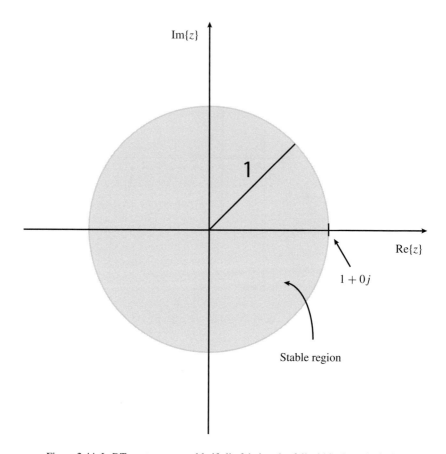

Figure 2.44 In DT, systems are stable if all of their poles fall within the unit circle.

large values of k. Notice that this differs significantly from the CT of two open-loop poles on the real axis in the LHP. In the CT case, the system would technically be stable no matter how large k became. The damping ratio would get increasingly small, and judging from the time-domain step response the system would appear to increasingly teeter on the edge of instability. But the fact would remain that the poles would never cross the all-important $j\omega$ axis.

2.5.5 A Useful Limit of DT

As we leave DT root locus for purposes of this book, it is useful to make an observation by way of establishing a unification of CT and DT methods. Suppose that we view CT as a limiting case of DT, which is to say that the

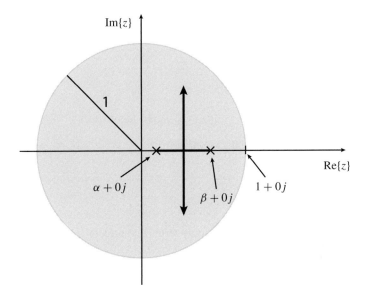

Figure 2.45 A root locus plot for example DT system with an overlay of the unit circle.

"time" increment is extremely small compared to the time scales of all poles, zeros, and, if relevant, signals, of interest. Start with a sampled CT exponential for illustration,

$$x(t) = e^{-\sigma_0 t} e^{j\omega_0 t} \tag{2.69}$$

$$x[n] = e^{-\sigma_0 n \cdot \Delta t} e^{j\omega_0 n \cdot \Delta t}.$$

Suppose that $x[n]$ as detailed in Eq. 2.69 is a term in the impulse response of a DT system, which means that pole of the system is therefore

$$p = r e^{j\Omega} \tag{2.70}$$

$$= \left(e^{-\sigma_0 \cdot \Delta t}\right) e^{j\omega_0 \cdot \Delta t}.$$

That is, for this DT pole we have $r = e^{-\sigma_0 \cdot \Delta t}$ and $\Omega = \omega_0 \cdot \Delta t$. Now the limit that we are interested in is for the time increment Δt to be extremely small compared to the time scale of the impulse response: the index n, say, must undergo one million steps before completing a full period of the oscillatory part.[13] This is the equivalent of stipulating both

$$|\omega_0 \cdot \Delta t| \ll 1 \qquad (2.71)$$

and

$$|\sigma_0 \cdot \Delta t| \ll 1. \qquad (2.72)$$

In this limit, we can step toward rewriting Eq. 2.70 as a first-order approximation:

$$p = r e^{j\Omega} \qquad (2.73)$$
$$\approx (1 - \sigma_0 \cdot \Delta t)(1 + j\omega_0 \cdot \Delta t).$$

Multiplying through, and keeping only terms that are constant or first-order in Δt, we get

$$p \approx (1 + 0j) + (-\sigma_0 \cdot \Delta t + j\omega_0 \cdot \Delta t). \qquad (2.74)$$

In this "oversampled" limit for DT systems, then, it is as though we "zoom in" to the portion of the unit circle in the vicinity of $1 + 0j = 1 \cdot e^{j0}$ as shown in Figure 2.46. Zoomed in like this, we cannot see the mighty arc of the unit

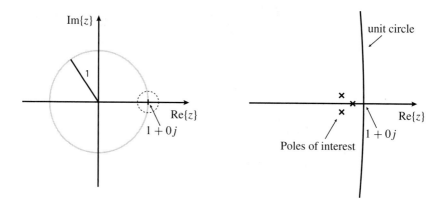

Figure 2.46 DT in the oversampled limit.

circle; to us instead the arc appears more like a long, straight line very similar to the $j\omega$ axis of continuous time. Indeed, the $1 + 0j$ point looks and acts like

---

[13] Our signal processing friends would call this *oversampling*. In contrast, sampling at the Nyquist rate requires only that the index undergo two or more steps before completing a full period of the oscillatory part.

the origin in CT systems: poles to left of this "origin" decay exponentially; poles with an imaginary part exhibit oscillatory behavior.

When we are *not* in this oversampled limit, a lot of weird stuff happens which is alien to those whose introduction to LTI theory comes through CT systems. The phenomenon of aliasing, for example, has no parallel in CT signal processing, and there is no such thing as a maximum frequency sinusoid (corresponding to $\Omega = \pi$). Between this odd behavior, and the disorienting change in notation going from s to z, a reasonable response for a student is to strictly segment CT and DT methods into different parts of their brain. True mastery, however, lies in seeing the common foundation despite the superficial differences. This oversampled limit is a good way to begin bridging the gap between CT and DT.

2.6 Common Control Strategies

Root locus techniques are a great way to evaluate feedback control strategies. When we speak of a "control strategy," what we mean is the designing of the dashed box in Figure 2.24. In this figure, we often are given $G(s)$ in the form of the thing we are trying to control (or, the "plant"), and $H(s)$ in the form of sensors wherewith we measure how the plant is actually behaving.

At this point is appropriate to introduce a few common control techniques. They are simple, they work well, and with just this handful of techniques you can get quite far in the design of single-input, single-output feedback systems.

2.6.1 Gain Reduction

Gain reduction to stabilize a feedback system is the absolute simplest thing you can do, and the truth is that in engineering simplicity should never be cast aside lightly. The experienced engineer cultivates simplicity, and parts with it only with great reluctance.

As a practical matter, almost everyone has experience with gain reduction from everyday life. Very often at a wedding reception or some other celebratory occasion, the host will approach an amplified microphone to make remarks to the guests. The signal from the microphone is greatly amplified and fed to the speaker system. These speakers, in turn, are acoustically coupled back to the microphone, completing what we now understand to be a feedback loop. Not only that, it is a feedback loop with *delay* in the loop transmission. The speed of sound is 343 m/s so if, say, the microphone is 10 m from the speakers, the delay is 29 ms, which is a massive phase shift at audio frequencies.[14] We all know what happens next: a tremendous, out-of-control howling from the speakers, many guests cover their ears and, interestingly, a few people shout, "Turn down the volume on the speakers!" That is to say, "Reduce the gain in the feedback loop!"

Returning to our more familiar block diagrams, it is not hard to imagine a situation in which gain reduction might help. We have argued that the behavior of many common systems is "dominated" by one or two poles. That does not mean, however, that other, faster poles do not exist. In fact, these faster poles exert a strong influence on performance if we pursue our feedback strategy with too much abandon.

As an example, suppose we had a plant that we had characterized as having two poles as shown in Figure 2.47. Based on this simple model, we expect

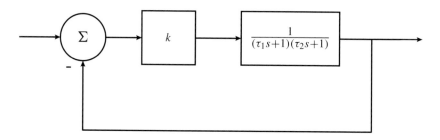

Figure 2.47 Example for the gain reduction strategy.

the root locus to be on the left of Figure 2.48. However, even one unmodeled pole somewhere out to the left on the real axis means the root locus in real life behaves like the right side of Figure 2.48.

In the laboratory, the result of this unmodeled pole is that our closed-loop poles are in the RHP and so: our robot shakes uncontrollably and tears itself apart; our RF amplifier ignores our input signal and slams like a zombie from supply rail to supply rail; our cruise control system becomes the subject of a nation-wide recall. A conservative strategy for eliminating these unhappy outcomes is that we reduce the gain parameter k.

Gain Reduction Is Bad

Always remember when you pull this trick, though, that it is a fairly radical thing to do. That is, *you are throwing away gain, and therefore immunity to variation in the forward path, at <u>all</u> frequencies*. Immunity to variation in the

[14] Check for yourself: "middle C" on the piano is commonly tuned to 261.63 Hz.

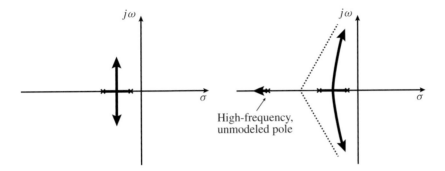

Figure 2.48 Based on our simple model of Figure 2.47, we expect the root locus on the left. However, it is normally true that our physical model is incomplete and that one or more high-frequency poles of the system are left out. This will mean that as we increase k, rather than simply a complex pole pair that is ever more lightly damped, we will have closed-loop poles actually enter the RHP as shown on the right.

forward path is the magic of feedback. There are going to be times when you decide it is the best thing to do, so don't as a rule ban gain reduction from your design practice. But there is usually a better way.

2.6.2 Dominant Poles and Integrators

A dominant pole is just what it sounds like: a low-frequency pole that is so slow that it "dominates" the dynamics of the forward path. It is a great way of simplifying life if the model of your plant has a lot of high-frequency stuff going on.

But why would a low-frequency pole ever be good? Sounds an awful lot like we're slowing things down, and slow is always bad, right?

Well, no. Remember the main crime of gain reduction, which was that it traded away precious gain at *all* frequencies. Qualitatively, it is the gain at low frequencies in a feedback loop that makes the system "dependable," "solid," and "repeatable," while the gain at high frequencies is qualitatively responsible for speed of response. So if we can't have everything, very often an excellent thing to do is to lock in the low-frequency feedback benefits by arranging for extremely high gain in that region. The dominant pole then makes sure that the loop gain falls below unity well before we get in the neighborhood of the fundamental oscillation condition.

An integrator in the forward path is a deservedly popular incarnation of the dominant pole compensation strategy. Remember this trick.

2.6.3 Lag and Lead Compensators

A good way to organize your thoughts on compensation is to prioritize frequency ranges for which you must have a lot of gain, versus those for which gain is unimportant. Dominant pole compensation is an example of prioritizing basically DC performance over everything else. Lag and lead compensators are the next step up in sophistication.

Roughly speaking, lag compensators are a method of prioritizing low frequency (not just DC) gain, while lead compensators are a way of prioritizing high-frequency gain, or speed of response. Rather than present these techniques in the abstract, it is helpful to imagine compensating a system that has two poles according to Figure 2.49. The Bode plot of the plant, and the root locus plot of the uncompensated system are as shown in Figure 2.50. You can see that while we will never technically get poles in the RHP, we will get unacceptable damping ratios if we crank up the gain too far. Given this as our starting point, what can be done with lag and lead compensators?

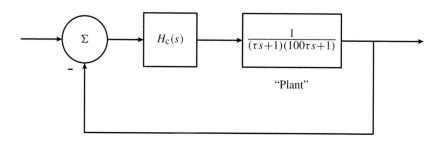

Figure 2.49 Example for lag and lead compensators.

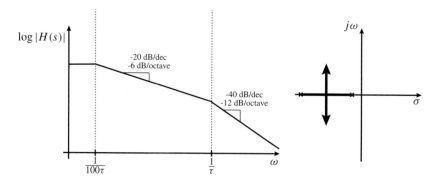

Figure 2.50 Bode plot asymptotes and root locus for the system of Figure 2.49.

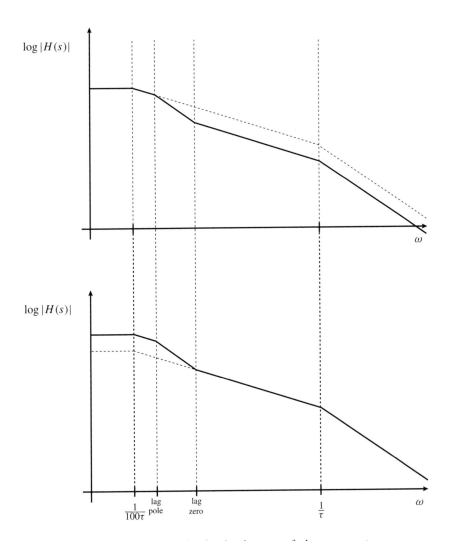

Figure 2.51 Bode plots showing the placement of a lag compensator.

A lag compensator is a pole-zero pair in which the pole is at a lower frequency than the zero. We can imagine applying it in this situation in one of two ways, shown in Figure 2.51. In the top plot, you can see that we have discarded gain at high frequencies (associated with fast speed of response) but done all we could to preserve gain at low frequencies. Resist the urge to make your control systems respond faster than they need to. Real systems almost always have unintended, high-frequency disturbances in the signal paths. A sufficiently fast system will react to every perturbation and will be noticeably "twitchy" or "jumpy." Don't say you weren't warned.[15] In the bottom graph, you can see that that we have used the lag to boost gain at low frequencies while leaving the high-frequency gain untouched.

To see the effect of a lag compensator on stability, study the Bode plots in Figure 2.60. You can see that the lag compensator allows us to do our

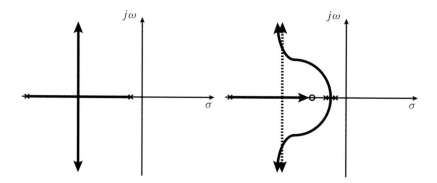

Figure 2.52 A root locus plot illustrating lag compensation.

horse trading of low- and high-frequency gain while still allowing for closed-loop poles with attractive damping ratios. A root locus plot illustrating lag compensation is shown in Figure 2.52.

In contrast to the lag compensator, the lead compensator is about speed. We use a lead compensator when we want to increase the gain at high frequencies, but for stability reasons are unable to also increase the gain everywhere else. A classic lead compensator looks like Figure 2.53. We see in the Bode plot that the low-frequency gain is left unchanged, but we have increased immunity to forward-path variations at high frequencies.

The root locus plot, in Figure 2.54, is also informative. We can see that even controlling for damping ratio of the dominant pole pair, the real part of the closed-loop complex pole pair moves to the left. We have thus sped up the system without compromising stability. The slower, real pole moves toward the zero for normal values of k, and the closer it gets to the zero, the less effect it has on closed-loop dynamics.[16]

---

[15] In electronics especially, it is easy to fall into the trap of thinking "fast" or "high bandwidth" is just inherently good. It is usually more complicated than that. If you want to hit one of your specifications out of the park, try lowering the *power consumption* of the systems that you build.
[16] Although if you look carefully at the closed-loop step response, you will often see a "long, slow tail" that seems to ride on top of an otherwise normal-looking step response.

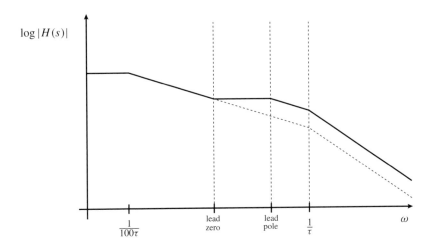

Figure 2.53 Bode plot for a lead compensator.

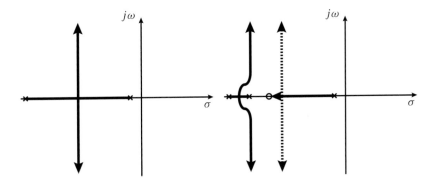

Figure 2.54 A root locus plot illustrating lead compensation.

2.6.4 PID Control

Taken together, gain reduction, dominant pole, lag-lead, and combinations thereof, form a very satisfying way of navigating the compensator design space. You can actually skip *this* subsection if you feel you have come to grips with Section 2.6 up to this point.

If you are still reading this subsection, it is probably because you have heard people talk about something called "PID" control. The short story is that PID, or proportional-integral-derivative control, is another way to organize one's thoughts to navigate the frequency-dependent compensator space. We might imagine a generic compensator with three adjustable parameters as shown in

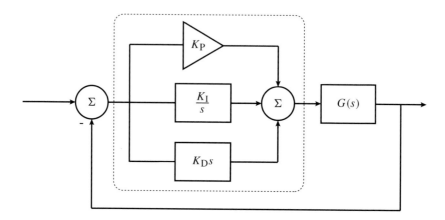

Figure 2.55 Proportional-integral-derivative (PID) control.

Figure 2.55, where K_P, K_I, and K_D determine just how much *proportional* (gain), *integral* (dominant pole), and *derivative* (strategically placed zero, perhaps for a lead or lag) to add to our compensator cocktail. Ultimately PID control still has to be about poles and zeros and root locus, of course, and so to see this we dive into a little bit of algebra:

$$H(s) = K_P + \frac{K_I}{s} + K_D s \qquad (2.75)$$
$$= \frac{K_D s^2 + K_P s + K_I}{s}.$$

Now, since we cannot build a pure differentiator, what with its infinite gain at infinite frequency, the real deal will look more like

$$H(s) = K_P + \frac{K_I}{s} + \frac{K_D s}{\tau_D s + 1} \qquad (2.76)$$
$$= \frac{(K_P \tau_D + K_D)s^2 + (K_P + K_I \tau_D)s + K_I}{s(\tau_D s + 1)}.$$

What we see is that with the three parameters K_P, K_I, and K_D, we are free to place a pair of zeros wherever we want, make a pole appear at the origin (or not, by making $K_I = 0$), and place one more real pole. With some work, you can map a discussion of PID control onto a discussion of gain reduction, dominant pole, and lead-lag compensation, and vice versa. It is best to study these methods, and decide for yourself which one works best for you. And if neither feel quite right, dive back into the concepts and come up with your own way of seeing things.

2.7 Answers to Sample Problems

Figure 2.56–2.60 are the answers to the problems posed in 2.5.2.

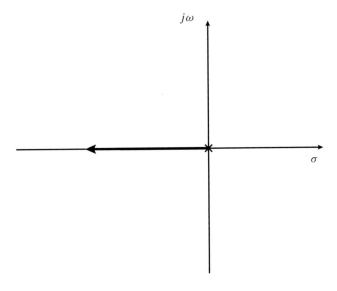

Figure 2.56 Example 1: $L(s) = \frac{k}{s}$.

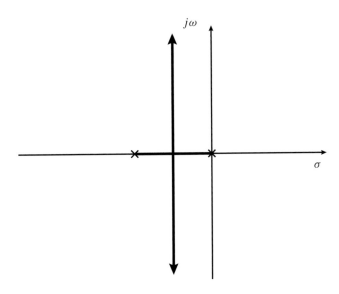

Figure 2.57 Example 2: $L(s) = \frac{k}{s(s+1)}$.

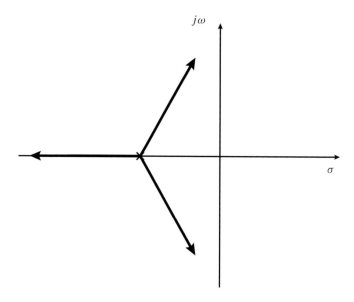

Figure 2.58 Example 3: $L(s) = \frac{k}{(s+1)^3}$.

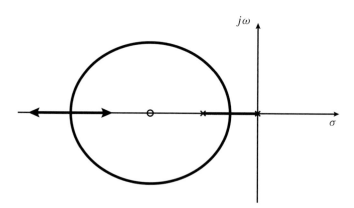

Figure 2.59 Example 4: $L(s) = \frac{k(s+2)}{s(s+1)}$.

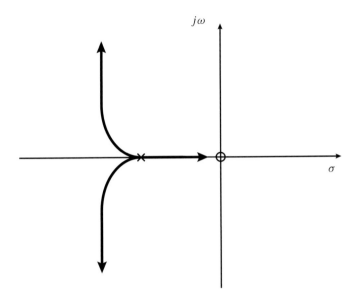

Figure 2.60 Example 5: $L(s) = \frac{ks}{(s+1)^3}$.

3

The Nyquist Stability Criterion

In this chapter we cover the Nyquist stability criterion: a much-feared, seldom-understood, devoutly avoided analysis technique among practitioners of the feedback art. If nothing else, it is worth understanding because it is the foundation of its much-beloved, seldom-understood, devoutly applied cousin: phase margin. Improper thinking about phase margin can fool you, but the Nyquist criterion never does. Let's unravel this.

3.1 An Authoritative Test of Stability

As good as root locus and transfer function based approaches are, we often find ourselves in situations where the transfer function in all its algebraic glory is not known. What is easy to measure, however, is the frequency response of the system we wish to control. That is, we pass a sinusoid of a known frequency through the block, note the resultant phase shift and amplitude scaling, and repeat for a closely spaced group of frequencies in the band of interest. Could it be true that given only the frequency response of a loop transmission, we have enough information to determine the stability of the closed-loop system? It sure seems like it ought to be possible. The Bode plot of the loop transmission is a complete characterization, after all.

It turns out that the answer is yes. In order to see this, we will have to take a brief detour through methods of complex analysis. The reward for your effort will be knowledge of a truly authoritative and comprehensive test of stability.

3.1.1 True Delay and Root Locus

Imagine a loop transmission of the following simple form:

$$L(s) = \frac{k}{s} e^{-sT_D}. \tag{3.1}$$

This is an integrator, followed by a *pure delay*. Pure delays are a common feature in feedback systems; the notorious microphone-too-close-to-the-speaker is a familiar example. This almost embarrassingly simple transfer function, however, falls squarely outside the analytical reach of root locus techniques. This is because we have no good way to capture e^{-sT_D} on a pole-zero plot.

So there you have it. We need another analysis technique if for no other reason than to handle pure delays.

3.2 A Note on Conformal Mapping

The Nyquist stability criterion requires the use of a complex analysis technique called conformal mapping. The concept is fairly simple[1] and is as follows.

We are used to working with the complex variable s. We are also used the complex valued function $F(\cdot)$, which takes variable s as an input and *maps s* to a new complex value $F(s)$. For a conformal map, we start with a contour in the s-plane. For every point s_0 on that contour, we evaluate $F(s_0)$ and get a new complex value. We then plot this new set of complex values on a separate set of axes.

That's it. That's conformal mapping. A few examples, now.

To understand Figure 3.1, it may be helpful to refer back to Section 1.5.2. For all points on this specially chosen contour in the s-plane, the double pole imparts a phase angle of $-45°$ ($= 2 \times 22.5°$). As we progress along the contour we get farther and farther away from the double pole, so the magnitude of $F(s)$ (which, in the $F(s)$-plane, is represented by the *distance from the origin*) decreases. The resulting conformally mapped contour is sketched on the right in the $F(s)$-plane of Figure 3.1.

The readers should place a heavy emphasis on the idea of "sketching," as opposed to "drawing accurately to scale." Notice that no values were calculated to determine, for example, exactly how far the mapped point A should be from the origin. Get used to this; when you are drawing conformal maps by hand, just as with root locus plots, you are a cartoonist, not a cartographer.

---

[1] Don't let most textbook treatments convince you otherwise.

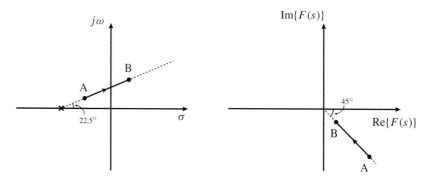

Figure 3.1 Conformal map example: $F(s) = \frac{1}{(s+1)^2}$.

It will normally be true that contours in the s-planes will cover many orders of magnitude in distance from the poles. It is impossible to draw these contours to scale on a set of linear axes. You will see that these cartoons nevertheless get across the salient features that we seek.

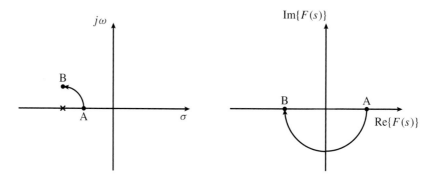

Figure 3.2 Another conformal map with $F(s) = \frac{1}{(s+1)^2}$, this time mapping a different contour.

In a second example, illustrated in Figure 3.2, notice that all of the points on the chosen contour in the s-plane just happen to be equidistant from the double pole. It follows that the corresponding mapped contour in the $F(s)$-plane will consist entirely of points that are equidistant from the origin. And so it is.

Figure 3.3 shows a slightly more complicated example, but following the guidelines of the first two examples gets you there. If you are new to conformal mapping, or rusty, a great exercise is to cover up the solution (the figure in the $F(s)$-plane) and try to reproduce it on your own with pencil and paper.

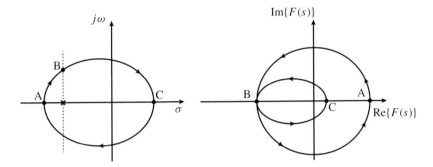

Figure 3.3 A third conformal map using $F(s) = \frac{1}{(s+1)^2}$, with a more complicated contour.

A classic mistake here is to simply follow along and feel like you "get it." Also, resist the temptation to go to computer aids to draw things for you at this point. Computers help the most when your conceptual understanding is firmly in hand. Invest the time now with pencil and paper, and you will be better with the computer when you use it.

3.3 Cauchy's Principle of the Argument

We need one more tool to completely set the table for the Nyquist stability criterion. That tool is called Cauchy's principle of the argument, a famous result from complex analysis.

It is simply stated.

Given a function $F(s)$ and a closed contour C in the s-plane such that $F(s)$ has no poles or zeros on C, then

$$N = Z - P, \tag{3.2}$$

where

N = no. of positive encirclements of the origin;

Z = no. of zeros of $F(s)$ inside C;

P = no. of poles of $F(s)$ inside C.

By "encirclements," we mean net trips a full $360°$ around the origin, and the positivity or negativity of such an encirclement is determined by the direction of the original contour in the s-plane. Really, the less said about this principle the better. Seeing a few examples is the best way of getting the idea.

The principle itself can be understood as a straightforward consequence of
Section 1.5.2.

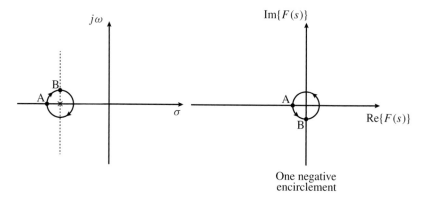

Figure 3.4 A conformal map using only a single pole, $F(s) = \frac{1}{(s+1)}$, showing
one *negative* encirclement of the origin in the $F(s)$-plane.

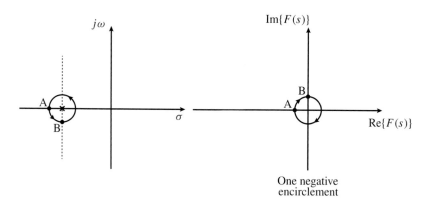

Figure 3.5 A conformal map using only a single pole, $F(s) = \frac{1}{(s+1)}$, with the
mapping contour going in the opposite direction as in Figure 3.4. This also results
in a negative encirclement, as it should.

A few examples are important to understanding encirclements. Note that
in both Figure 3.4 and Figure 3.5, the encirclement of the origin is the
opposite sense of the original contour C. When C encircled the pole in the
clockwise sense, the mapped contour encircled the origin in the counterclock-
wise sense, and vice versa. This is what we mean by a "negative" encirclement
for purposes of Cauchy's principle of the argument.

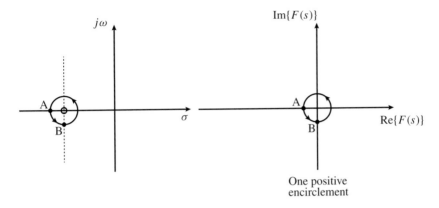

Figure 3.6 A conformal map using the mapping function $F(s) = s + 1$. Since we are circling a zero instead of a pole, we wind up with a positive encirclement of the origin in the mapped plane.

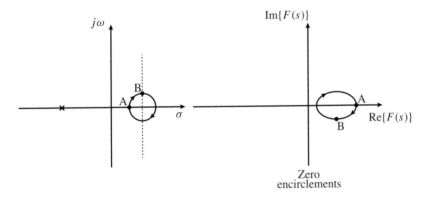

Figure 3.7 A conformal map using the mapping function $F(s) = \frac{1}{(s+1)}$. In this case, we have no net encirclements of the origin in the mapped plane.

There are three more examples in Figures 3.6, 3.7, and 3.8. If you can work through each of these examples on your own, starting with a blank sheet of paper, then you will have the basic idea. You can decide for yourself how proficient you want to become in problems such as these. Unless you are pressed to do so by a demanding application or you are a student facing an exam, exotic cases will come up only rarely.[2]

---

[2] Note that you can improve through repetition, but that repetition without thinking leads only to pattern matching. Take your time, and think through the steps.

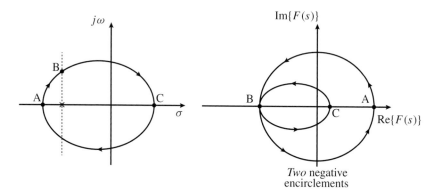

Figure 3.8 A conformal map using the mapping function $F(s) = \frac{1}{(s+1)^2}$. This mapping results in two net negative encirclements of the origin in the mapped plane.

3.4 And Now ... the Nyquist Stability Criterion

Okay, so now we are least novices with conformal mapping, and we have this neat result that the conformal map of a closed contour in the s-plane circles the origin if that closed contour encloses singularities like poles and zeros. How do we relate this tool to the problem of the stability?

We want to know if a system is stable. If that system has transfer function $\frac{L(s)}{1+L(s)}$, the question then becomes whether $F(s) = 1 + L(s)$ has any zeros in the right-half plane. Now, how can we use our shiny new tool, Cauchy's principle of the argument? We might begin by sketching a contour that encloses the whole of the right-half plane in Figure 3.9.[3] We could do a conformal map and count positive encirclements of the origin in the mapped plane. If the number of positive encirclements is greater than or equal to 1, then the system is unstable!

It is important to emphasize why this technique is useful as a practical matter. It is normally true in applications that we do *not* have a closed-form expression for the open-loop transfer function $L(s)$. However, we can usually measure $L(s)$ directly and enjoy the astounding result that these measured data, through the magic of conformal mapping and the Cauchy principle of the argument, can tell us whether $1 + L(s)$ has any right-half plane zeros, and if so, how many. This in turn tells us how many right-half plane *poles* our closed-loop system has, if any. This overall method is called the Nyquist test,

---

[3] Yes, the right-half plane is technically of infinite expanse. Remember: cartoonist, not cartographer.

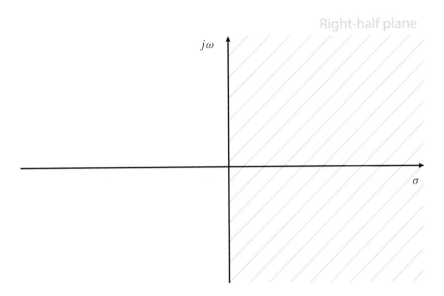

Figure 3.9 For the Nyquist stability criterion, we are looking for zeros of mapping function $F(s) = 1 + L(s)$ in the right-half plane.

or more formally, the Nyquist stability criterion. We will add some refinement in the coming paragraphs to make it easier to use, but for the first time you have the basic idea.

The sharp-eyed reader may point out that the accounting with the Cauchy principle expresses encirclements of the origin in terms of both the poles *and* zeros of $1 + L(s)$ within this all-RHP-encompassing contour. So if we're counting *net* encirclements, we might be fooled if we don't separately keep track of the RHP poles of $1 + L(s)$, which are the same as the poles of $L(s)$. And we just finished admitting that we don't normally have a closed-form expression for $L(s)$. Gotcha!

No problem, in practice. If we switch on our open-loop system, provide it no input, and the output races off exponentially in time toward one limit or another, then we know that we have one or more RHP poles in $L(s)$. If *not*, then we *don't*. And if we *do*, then we have no choice but to dig into the physics of the system to find out how many such poles there are. Admittedly, that can get complicated. But when the system is open-loop stable, as is typically the case, then we don't have to worry.[4]

---

[4] Real-world examples of open-loop unstable systems include inverted pendulums, and often magnetic levitation systems.

We still have the practical problem of choosing sensible contours. Such a contour should have at least two properties. First, we should be able to leverage knowledge of the measured frequency response $L(s)$ to do the mapping. Second, it should encompass the whole of the RHP.

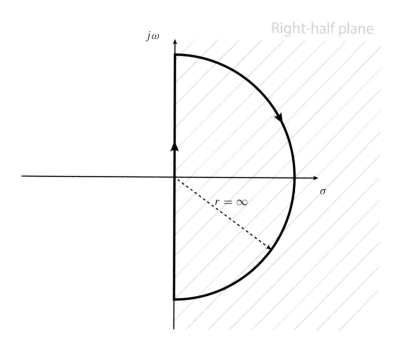

Figure 3.10 The "D contour," often used with the Nyquist stability criterion.

The contour in Figure 3.10 has both desirable properties. This is often called the "D contour." The flat part of the contour is the $j\omega$ axis; we can measure $L(j\omega)$, and just add 1 $(+0j)$ to every measurement point to map this part of the contour. The curved part of the contour seems like a problem, until we realize that every point on the contour is an infinite distance from the poles and zeros of $L(s)$. Provided $L(s)$ represents a physical system with more poles than zeros, $|L(s)| = 0$ in the limit $R \to \infty$. It follows that $1 + L(s)$ for all of the parts on the curve of the contour map to 1 in the $F(s)$-plane.

We continue to drive toward a strategy for determining the stability of a closed-loop feedback system based on frequency domain measurements of $L(s)$. We take the measured data and do a conformal map of the D contour using mapping function $F(s) = 1 + L(s)$. We count positive encirclements of the origin in the $F(s)$-plane and use Cauchy's principle of the argument.

There is at least one clever change we can make to make the application of this strategy criterion easier.[5] As it stands, we must take all of our frequency domain data, add $1 + 0j$ to every data piont, and plot the new set of complex points in what we are calling the $F(s)$-plane. We can save ourselves some work, though, by using $L(s)$ as the mapping function *directly*, taking advantage of the fact that $L(s)$ is very simply $F(s) - 1$. This means that our new mapped contour is the same as the old, only translated to the left by 1. All this means is that instead of counting encirclements of the origin, we now count encirclements of $-1 + 0j$.

One other change that we can make harkens back to our practice in root locus plots of keeping around a parameter as part of the loop transmission. That is, $L(s) = kL_0(s)$. If we express $L(s)$ in terms of our original mapping function, we have

$$L_0(s) = \frac{1}{k}(F(s) - 1). \tag{3.3}$$

Why bother? What this means is that we do not need to redraw the Nyquist plot every time we change the gain of the loop transmission. We can do *one* Nyquist plot and count encirclements of the $-\frac{1}{k} + 0j$ point in the mapped plane, where k is the gain parameter. It is an important tweak to the technique. If it's not clear why now, it will become clear the first time you labor through the Nyquist test and then wonder what happens if you change the gain.

So there you have it! This is the Nyquist stability test. To summarize:

1. Draw a "D" contour in the complex plane.
2. Evaluate $L_0(s)$ at every point on the D contour, and plot these new mapped points on the $L_0(s)$-plane.
3. Count encirclements of the the the $-\frac{1}{k} + 0j$ point, and apply the Cauchy principle of the argument to determine how many zeros of $1 + L(s)$ are in the RHP. If the answer is one or more, your system is not stable.

A simple example is in order. Consider applying our new methodology to the single-pole loop transmission $L_0(s) = \frac{1}{(s+1)}$, illustrated in Figure 3.11. In this figure, you can see that the positive $j\omega$ axis maps to the lower half of of the conformal map; the negative $j\omega$ axis maps to the upper half of the conformal

---

[5] If this is your first trip through the Nyquist criterion, you might be impatient at this point over making a long development even longer in service of "making things easier." Hang in there. If we skipped these last refinements, applying the Nyquist test by hand would quickly get needlessly, annoyingly tedious.

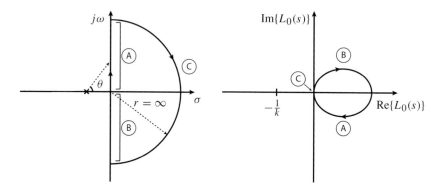

Figure 3.11 Applying the Nyquist stability test to $L(s) = \frac{1}{s+1}$. Note that the entire curved part of the D contour maps to the origin in the $L(s)$-plane, because $r = \infty$. This example illustrates, for positive k, that there are no *zeros* of $1 + L(s)$ in the RHP.

map; the curved part C of the D contour maps entirely to the origin in the $L(s)$-plane. For positive frequencies, near the origin the magnitude of $L(s)$ is close to unity. As we sweep out to high positive frequencies, the magnitude shrinks to zero and the phase contributed by the pole approaches $90°$. We follow a similar mapping for negative frequencies. Overall, we are checking for positive encirclements of the $-\frac{1}{k} + 0j$ because we are trying to determine if there are any *zeros* of $1 + L(s)$ in the RHP.[6]

3.5 Bode Plots Help with Nyquist

One of the skills that is drilled into almost everyone who studies linear system theory is how to draw a Bode plot. Now is a great time for you to review that skill, especially if you are rusty. The methodology can easily be mapped to our geometric understanding of poles and zeros as detailed in Section 1.5.2. The reason that Bode plots are often broken out as a special topic is that, as it happens, they can often be sketched with great accuracy using only pencil and paper. It helps that the vertical scales for the magnitude plots are logarithmic, and also that poles and zeros in nature are quite often separated by orders of magnitude. For a variety of reasons, facility with Bode plots winds up being extremely useful in engineering practice.

---

[6] The emphasis and repetition here is because at some point nearly everyone confuses themselves on poles and zeros with the Nyquist test. We are looking to see if there are closed-loop poles of $\frac{L(s)}{1+L(s)}$ in the RHP. This means hunting for *zeros* of $1 + L(s)$ in the RHP.

We bring this up now in connection with drawing Nyquist contours based on the frequency response $L(s)$. Remembering that the flat part of the D-contour is just the $j\omega$ axis, then our mapping function $L(s)$ is nothing more than the frequency response $L(j\omega)$. It is useful to see an example of a Bode plot directly leveraged into a Nyquist plot.

Consider the open-loop transfer function $L(s)$:

$$L(s) = \frac{100}{(s+1)(10^{-3}s+1)}. \qquad (3.4)$$

The Bode plot looks as shown in Figure 3.12. With practice, you can render Nyquist plots from Bode plots as follows. Moving from left to right,

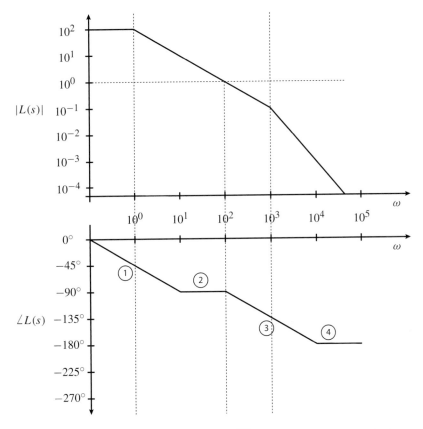

Figure 3.12 A Bode plot for $L(s) = \frac{100}{(s+1)(10^{-3}s+1)}$. Combining the magnitude and phase plots in your mind will help you to synthesize the corresponding Nyquist plot.

and following the circled numbers in the Bode plot, we simply render the qualitative features of the Bode plot into the Nyquist plot:

1. Figure 3.13: The magnitude declines slightly, and the pole causes the phase to sweep toward $-90°$.
2. Figure 3.14: We've passed the first pole, so the contour should now be tracking steadily in toward the origin. The phase is at about $-90°$.
3. Figure 3.15: We've hit another pole. We continue tracking toward the origin, but begin sweeping toward our final angle of $-180°$.
4. Figure 3.16: We asymptotically approach $-180°$ as we bury ourselves into the origin.

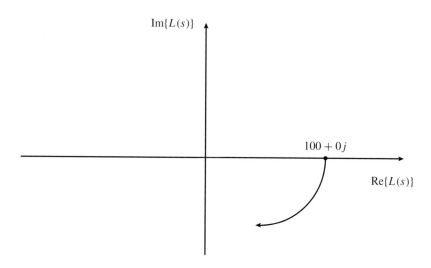

Figure 3.13 The magnitude declines slightly, and the pole causes the phase to sweep toward $-90°$.

Note finally that we have only mapped the positive $j\omega$ axis of our D-contour. To map the other half, we exploit a known symmetry of $L(s)$. Assuming that $L(s)$ is the Laplace transform of a purely real impulse response, we know that $L(s)$ is *conjugate symmetric*: $\angle L(j\omega) = -\angle L(-j\omega)$, and $|L(j\omega)| = |L(-j\omega)|$. Take all the time you need to convince yourself that this means we can complete the Nyquist contour as shown in Figure 3.17.

In the next section, we will use Nyquist plots to evaluate the stability with this $L(s)$ as its forward path. Remember, too, that, as with root locus, we are interested in the thought experiment of knowing what would happen if we were to fiddle with the gain. That is, the actual system to keep in mind is in Figure 3.18.

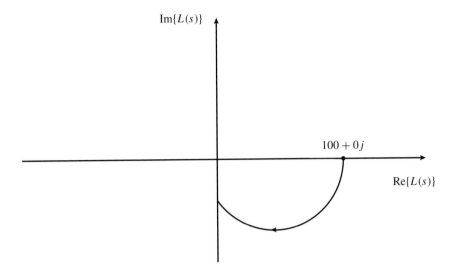

Figure 3.14 We've passed the first pole, so the contour should now be tracking steadily toward the origin. The phase is at about $-90°$.

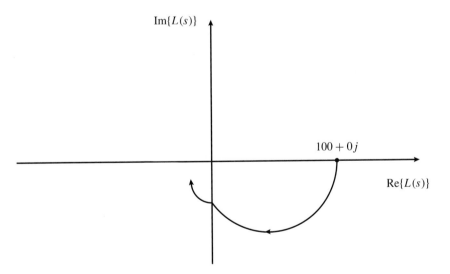

Figure 3.15 We've hit another pole. We continue tracking toward the origin but begin sweeping toward our final angle of $-180°$.

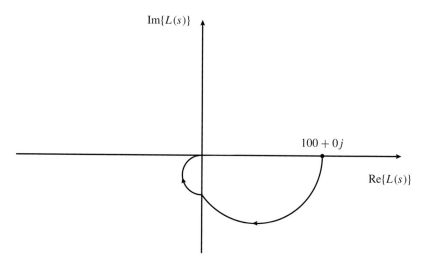

Figure 3.16 We asymptotically approach $-180°$ as we bury ourselves into the origin.

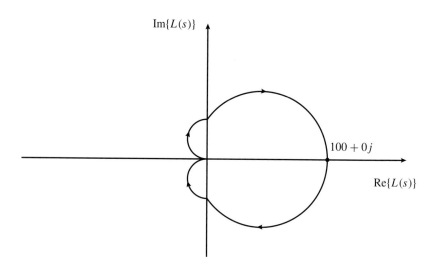

Figure 3.17 Using the Bode plots in Figure 3.12 as a guide, we arrive at the complete Nyquist plot for $L(s) = \frac{100}{(s+1)(10^{-3}s+1)}$.

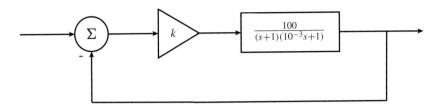

Figure 3.18 Parameterized block diagram upon which we will perform our Nyquist stability analysis. The loop transmission remains $L(s) = \frac{100}{(s+1)(10^{-3}s+1)}$, but with a multiplicative free parameter k.

3.6 Nyquist Plot Examples

Now for a couple of examples. For the first example, we can continue working with

$$L(s) = \frac{100}{(s+1)(10^{-3}s+1)},\qquad(3.5)$$

since we've already gone to the trouble of doing the Nyquist contour. We redraw the contour in Figure 3.19 and add the $-\frac{1}{k} + 0j$ point, whose encirclements we intend to chronicle.

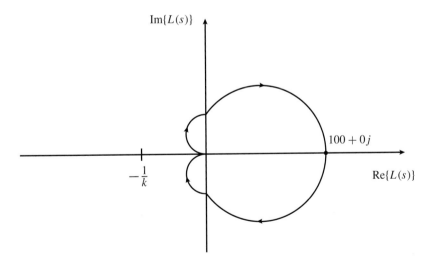

Figure 3.19 The Nyquist plot of Figure 3.17, but with a $-\frac{1}{k} + 0j$ point added.

Now, we set up a system of accounting. We know that

$$Z = N + P, \tag{3.6}$$

where Z is the number of RHP zeros of $1 + L(s)$, P is the number of RHP poles of $L(s)$, and N is the number of *positive* encirclements (clockwise in this case, since we went *upward* on the $j\omega$ axis when traversing the D-contour. This is important; make sure you understand this sign convention). We want Z because the RHP zeros of $1 + L(s)$ are the RHP *poles* of the closed-loop transfer function. If $Z > 0$, then our feedback system is unstable.

Begin the accounting. We know in this case that $L(s)$ has no RHP poles, so P is always zero. It helps to construct a table, such as in Table 3.1. For our

Table 3.1 *Ranges of k and resulting stability determination*

Range of k	P	N	$Z(=N+P)$	Stable?
$\infty < k < -0.01$	0	1	1	No!
$-0.01 < k < 0$	0	0	0	Yes
$0 < k < \infty$	0	0	0	Yes

more visual learners, we can redraw the Nyquist contour and label ranges for k as in Figure 3.20. Interestingly, there is a region where we have *positive* feedback, e.g., $k < 0$, yet $|k|$ is too small to actually cause instability. We can check this against the root locus plot in Figure 3.21. In this figure, we show the locus for positive values of k (which, in our convention, corresponds to *negative* feedback). We conclude that no matter how big k gets, we never actually get RHP poles in the closed-loop transfer function. It is true that for very high values of k the pole pair would be so lightly damped as to be unusable, but in the strictest sense, that is not unstable.

For negative values of k (corresponding, remember, to *positive* feedback) the root locus looks like Figure 3.22. From this figure, we would conclude that for small magnitudes of k, both real poles remain in the LHP. If we push it and make the magnitude of k too big, we wind up with a real pole in the RHP and all the trouble that causes. But we have the correspondence we sought: the Nyquist method and the root locus method both yield the same qualitative insights. Notice that in the case of Nyquist, we got *quantitative* values for k for different stability outcomes, which is new and thrilling.

Now suppose, as often happens, that the transfer function of this last example turned out to be only mostly accurate, and the truth is that we actually have

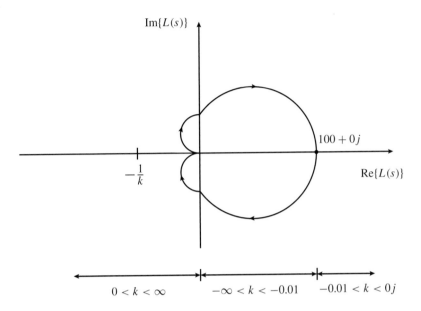

Figure 3.20 The Nyquist plot of Figure 3.17 that allows for determining stability for various values of k.

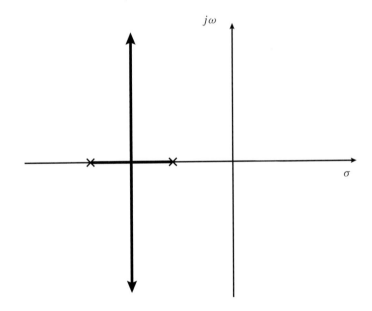

Figure 3.21 Checking Nyquist analysis with root locus. According to Figure 3.20, there will be no closed-loop RHP poles for positive values of k. We arrive at the same conclusion through root locus.

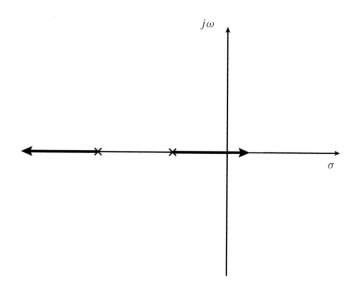

Figure 3.22 Checking Nyquist analysis with root locus. According to Figure 3.20, there will be no closed-loop RHP poles even for negative values of k, at least until the magnitude of k becomes great enough. Again, Nyquist and root locus analysis are in agreement.

some high-frequency poles that are unmodeled.[7] With closed-loop systems we must be careful. We can see this if we reanalyze that last example. This time, we add a third, previously unmodeled pole at 10^4rps. That is to say, now

$$L(s) = \frac{100}{(s+1)(10^{-3}s+1)(10^{-4}s+1)}. \tag{3.7}$$

To aid in drawing the Nyquist contour, we start with the Bode plot as shown in Figure 3.23. Following the same procedure as before, we can draw the half of the Nyquist contour that corresponds to the positive $j\omega$ axis, shown now in Figure 3.24. We then complete the contour by exploiting the conjugate symmetry of $L(s)$. The complete contour is shown in Figure 3.25.

Now what happens when we do the accounting? It looks like for $k = 1$, which is just the unmodified loop transmission $L(s)$, the system is safely stable. But whereas before, we were free to make k as large a positive value as we wished, now it appears that we encounter trouble once k gets to 10^3 or so. The new accounting is shown in Table 3.2.

[7] This is almost always the case; we should never forget that the we employ mathematical models because they are often *useful*, though never *complete*. Also, now is a good time to remind yourself of why linear systems' behavior is "dominated" by the low-frequency poles as shown in Section 2.5.3.

Table 3.2 *Ranges of k and resulting stability determination,*
now with a previously unmodeled pole

Range of k	P	N	$Z(= N + P)$	Stable?
$\infty < k < -0.01$	0	1	1	No
$-0.01 < k < 0$	0	0	0	Yes
$0 < k < \infty$	0	0	0	Yes
$10^3 < k < \infty$	0	1	1	No

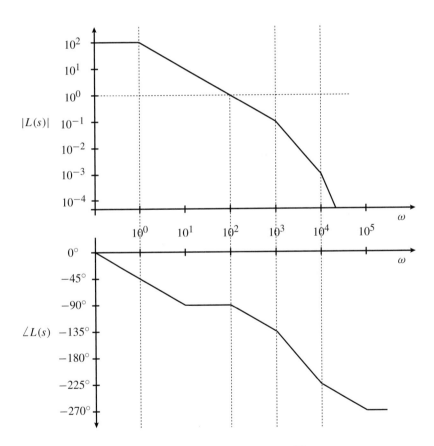

Figure 3.23 A Bode plot for $L(s) = \frac{100}{(s+1)(10^{-3}s+1)(10^{-4}s+1)}$.

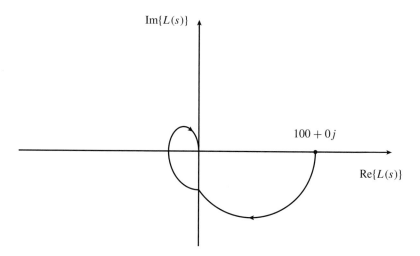

Figure 3.24 Half of the Nyquist plot for $L(s) = \frac{100}{(s+1)(10^{-3}s+1)(10^{-4}s+1)}$, which corresponds to the positive $j\omega$ axis from the D-contour.

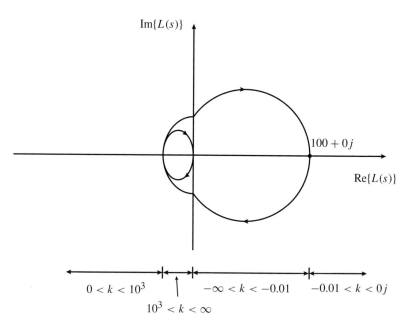

Figure 3.25 The full Nyquist plot for $L(s) = \frac{100}{(s+1)(10^{-3}s+1)(10^{-4}s+1)}$. The accounting for ranges of k corresponds to Table 3.2.

A takeaway here is that when it comes to design of feedback systems, don't get greedy on bandwidth and/or speed of response. The more aggressive you get, the more likely you are to encounter dynamics of your model that are not easy to account for.

With that, congratulations! If you have read this far, and worked to understand the material, you have the basic idea behind a foundational stability analysis technique. Do not expect yourself to be proficient at Nyquist plots yet, however. There are important special cases that come up.[8] Still, if you choose to pursue proficiency, that path will be made easier having grounded yourself in the concepts here.

As a final note, it worth emphasizing that while Nyquist and root locus are both definitive stability tests, root locus is somewhat more limited since it cannot treat pure delay, and because it requires closed-form expressions for the transfer functions. What we are about to tackle in the next section, however, is not definitive. Phase margin is handy: a rule of thumb that is useful and accurate and informative in a wide variety of real-world applications.[9] Use it freely according to its *strict* definition. But when you seek to build that symbiosis between analysis and conceptual understanding, let Nyquist and/or root locus play the part of your analytical tool.

3.7 Phase Margin: Why You Never Really Learned Nyquist

The Nyquist stability criterion is a comprehensive analytical tool that has a lot to recommend it. It is unambiguous, and it tells you for sure whether the system is stable. Maybe its finest attribute is that one can apply it using measured frequency-domain data without having to generate an equation-based model. The only downside? Applying it is a lot of work!

And that matters when doing design. When deeply engaged in an especially iterative design process, it makes all the difference in the world whether the impact of a design tweak can be assessed in seconds or minutes versus hours or longer. When changes can be evaluated quickly, a designer feels driven to optimize for for the absolute best performance, and is more likely able to navigate the design space with ease, confidence, and purpose. When the tweaking process is onerous, a psychological wall can build up against

---

[8] For example, the Nyquist contour should not hit any singularities, so what does one do for the case of an integrator, which is an open-loop pole at the origin? See how to handle this particular case in Section 4.2.

[9] That is to say, almost all situations that are not PhD qualifying exams.

changing anything about the system for any reason, especially once the system appears to be working reasonably well.[10]

With the Nyquist test, what we often do in real life is substitute for it with a measure called *phase margin*. This is so much easier to use, and so much faster to check, and so commonly gives the correct result that, well, it winds up being the technique that many designers use and remember. Let's revisit.

3.7.1 The Stability Margin Concept

The stability margin concept arises from the relative lack of nuance in the discussions of root locus and Nyquist plots. Reading about these techniques, you might get the mistaken impression that we limit our stability discussion to stable or unstable, good or bad.

In practical application, we are interested in nuance. Why care about "how stable" a system is? The answer is that *degrees of stability* typically trade against other performance metrics that we care about.[11] Consider a cruise control system, in which a "step" input might be a command to go from 55 mph to 60 mph. Figure 3.26 is how, as a driver, you expect the the car to react. If the control system has poles that are close to the $j\omega$ axis, however, you have the disaster depicted in Figure 3.27. As we adjust the parameters of the control system, there is likely to be a spectrum of behaviors as shown in Figure 3.28.

Thus the concept of stability margin is born. We ask, If the system is stable, how "close" is it to being unstable? If the closed-loop poles are in the LHP, are they deep in the LHP, or are they just shy of the $j\omega$ axis? If there were no positive encirclements of the $-1/k$ point for the Nyquist plot, did the contours "almost" circle it? It so happens that for all of these stability tests, coming close to being unstable corresponds to step response behavior that is close to oscillatory.

It turns out that "phase margin" is how we formally characterize degree of stability in the case of Nyquist plots. The beauty of phase margin, we shall see, is that you can measure it on a Bode plot and save yourself the whole trouble of doing a proper Nyquist plot. You should just be aware that this is a shortcut and that you should avoid generalizing from the phase margin definition for conceptual insight. In fact, it might be worth it even now to take a peek at Section 4.2 to make this cautionary note a little more concrete.

---

[10] Of course, one hazard of having the ability to do fast tweaking is neurotic, mindless, blind iterating devoid of insight into the trade-offs being made. Avoid.

[11] An extremely common example is that safely stable systems often have a slower speed of response, and/or a greater tolerance for component variations, than systems with a fast speed of response.

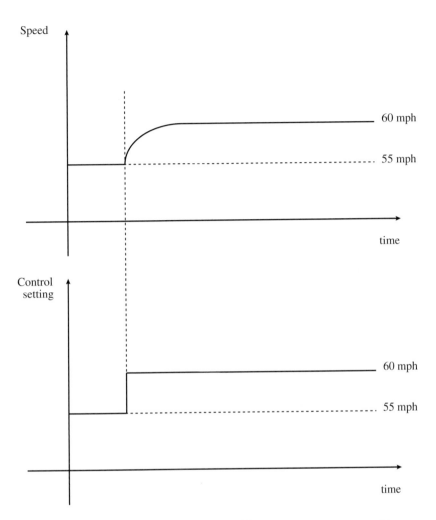

Figure 3.26 How we expect the cruise control in a car to react to a command to go from 55 to 60 mph.

3.7.2 Phase Margin Definition

On a Nyquist plot, we know that we encounter instability if the Nyquist plot encircles the $-1 + 0j$ point. In characterizing the *phase margin*, by which we mean, "How close do we come to encircling the $-1 + 0j$ point," we look at the plot as shown in Figure 3.29. (Stare at this plot until you get it. If you are stuck and remain so, revisit after you have read the rest of this chapter.)

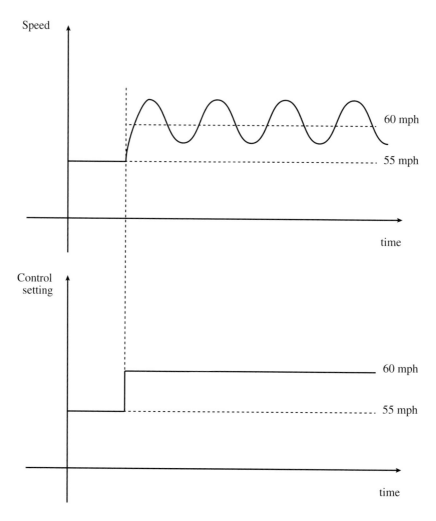

Figure 3.27 How we *do not* expect the cruise control in a car to react to a command to go from 55 to 60 mph.

If you are having trouble at this point, it is probably because you remember the concept of phase margin having to do with "how much phase you have to add at the point where the magnitude crosses unity until you get to −180°". In most treatments, this reasoning is usually reinforced by a Bode plot.

We are not going to do that here, because the whole thought experiment of "adding phase" at a single frequency is so needlessly and gratuitously divorced from physical reality. The author does not object to such divorces as a matter

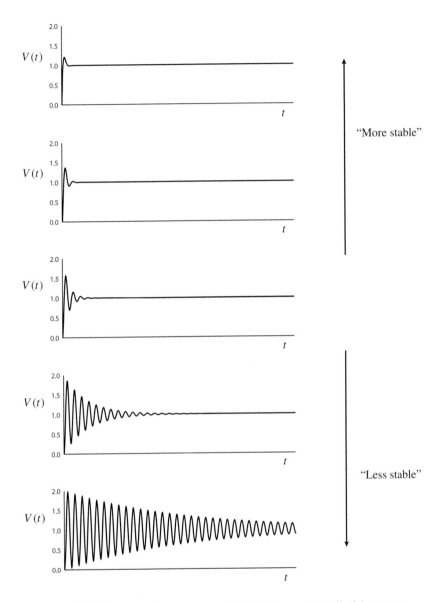

Figure 3.28 How we arrive at the concept of stability margins. All of these step responses are indicative of dominant pole pairs solidly in the left-half plane and therefore of stable systems. But it is hard to resist thinking of the bottom system as being "closer" to instability than the top system. Don't resist.

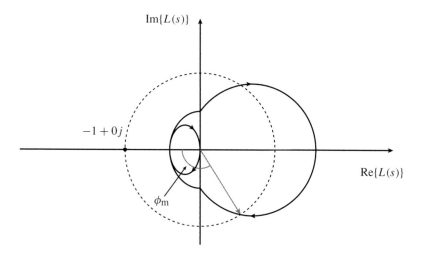

Figure 3.29 Phase margin: "When the Nyquist contour crosses the unity magnitude circle, how much more negative phase would be needed to barely encircle the $-1 + 0j$ point?"

of principle, but has observed the rapid descent from such innocent sloppiness to the absolute misunderstanding described in Section 4.2. Let's stick with Nyquist plots for the moment.

To better understand this idea of "adding phase," we should step back and remember that there is no physical way to add phase at one frequency and leave the rest of the Nyquist contour untouched.[12] Our choices for adding phase are actually quite limited: we can either add poles and/or zeros to the transfer function, or add pure delay. That's the list. The first option is complicated by the fact that poles and zeros affect not just the phase, but also the magnitude response. It follows that when adding a pole and/or a zero, the frequency at which the transfer function reaches unity must change unless a corresponding scalar gain also modifies the transfer function. But pure delay, on the other hand, is somewhat simpler, as seen in the the Bode plot for $D(s) = e^{-sT}$ shown in Figure 3.30. So if we modify a loop transmission $L(s)$ by adding delay $D(s)$, then we affect points on the Nyquist curve corresponding to low frequencies hardly at all, while at high frequencies we induce a veritable death spiral.[13]

---

[12] And try to imagine how odd a contour would look under such a modification.

[13] Here, "low frequencies" are those for which $\omega \ll \frac{1}{T}$. Don't breeze through this distinction, but take however much time you need to absorb it. Whether delay has an impact totally depends on the time scale that matters to you.

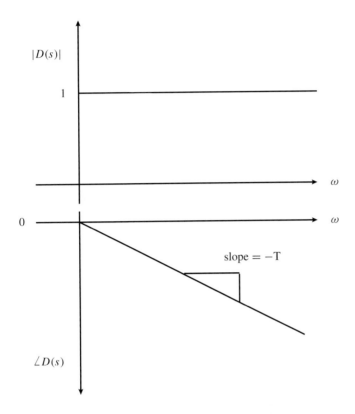

Figure 3.30 Bode plot for a $D(s)$, a pure delay.

Let's go back to our previous example and add a delay such that at the frequency at which $|L(j\omega)| = 1|$, the phase of the delay is exactly $-\phi_m$ as denoted on the Nyquist plot. Convince yourself that the new Nyquist plot would look like the sketch in Figure 3.31. *Now* it makes sense. Adding a delay whose phase at the frequency ω_c for which $|L(j\omega_c)| = 1$ would cause the phase of $L(j\omega_c)D(j\omega_c)$ to be $-180°$ would *indeed* push the $-1 + 0j$ point to the brink of encirclement.[14]

Fine. We have been extremely careful in our construction here, and it may seem that we have been overly so. At the end we find that we need not modify the definition of phase margin that is given in any textbook on feedback theory, which is

---

[14] Once we have the phase to $-(180° + \epsilon)$ where ϵ is an infinitesimal, negative offset, we go from "the brink of encirclement" to "encirclement."

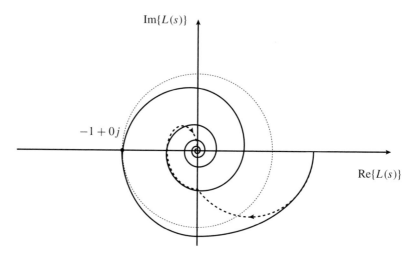

Figure 3.31 The half Nyquist plot of Figure 3.24, but modified with delay term $D(s)$. Notice that for low frequencies, where the delay does not add much phase, the Nyquist contour is unchanged. But if the delay is chosen to drive the phase margin exactly to zero, the result is that the $-1+0j$ intersects the Nyquist contour. This becomes a full-blown encirclement for even the tiniest bit of additional delay.

- **Phase margin:** If ω_c is the "unity-gain crossover frequency," where $|L(j\omega_c)| = 1$, then the phase margin is the difference $\phi_m = \angle L(j\omega_c) - (-180°)$.

Conveniently, the phase margin can be read directly off of a Bode plot as shown in Figure 3.32. But a major point of this section is not to get lazy when thinking about phase margin. It is very easy to look at a Bode plot and slip into a mode of thinking where a phase, disembodied from physical reality, is somehow "added" to a point on the transfer function where as if through magic it creates or causes an oscillation. Always remember that the phase margin trick originates in the Nyquist criterion, and it will create a much stronger foundation for your understanding.

Finally, be careful not to overinterpret phase margin. In particular, it is tempting to say that if it is bad for the loop transmission to have a gain of unity when the phase is $-180°$, it must be so much worse if the gain *exceeds* unity when the phase is $-180°$. Section 4.2 goes through an instructive example of how this type of thinking can go wrong. Just remember to use phase margin as helpful shortcut only. When you seek physical insight,

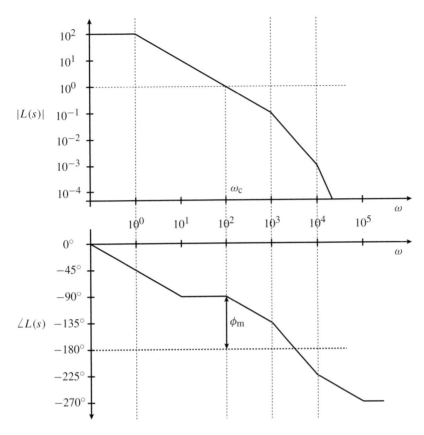

Figure 3.32 Reading phase margin off of a Bode plot of the loop transmission $L(s)$. Variable ω_c is often called the "unity-gain crossover frequency," or just "the crossover frequency."

use the other analytical tools in your bag like Nyquist stability, root locus, and the underlying differential equations that capture the physics of your system.

3.7.3 Phase Margin, Overshoot, Ringing, and Magnitude Peaking

We now refer back to Section 2.5.3, where we justified the empirical observation that many systems behave as though they are dominated by a single complex pole pair. In practice, it is useful to have a relationship between phase margin and the characteristics of second-order systems.

The following useful rules, along with justifications and cautionary notes, can be found in Thomas Lee's *The Design of CMOS Radio-Frequency Integrated Circuits*.

- Damping ratio as a function of phase margin:

$$\zeta \approx \frac{\phi_m}{100} \tag{3.8}$$

  with ϕ_m expressed in degrees.
- Peak overshoot in the time-domain step response:

$$P_o \approx 1 + \exp\left(\frac{-\pi\phi_m}{\sqrt{10^4 - \phi_{m^2}}}\right). \tag{3.9}$$

- Resonant peaking in the magnitude part of the Bode plot:

$$M_p \approx \frac{1}{\sin\phi_m}. \tag{3.10}$$

The phase margin here is always expressed in degrees, as opposed to radians. For many commonly encountered systems and phase margins, these relationships will be good to rapidly estimate performance.

3.8 Nyquist and Bode Techniques for DT Systems

It is possible to run through all of the foregoing treatment for DT systems, working out analogous frameworks. In practice, however, one tends not to need these techniques. In DT systems, the exact pole locations are usually known. Using computational aids, it is typically straightforward to do extremely accurate root locus plots. The result is that while one can work out Nyquist and Bode techniques in gory detail, this is typically done for analytical completeness and not as a practical aid.

4

Some Common Loose Ends

After a brief trip through the basics of LTI systems in Chapter 1, we covered the conceptual heart of feedback theory in Chapters 2 and 3. The intention was that these chapters would be easy to follow and provide a satisfying connection between mathematical formulations and conceptual, intuitive understanding. While many specialized topics have been omitted, material of Chapters 2 and 3 formed the backbone of a feedback systems course that was taught for many years at the Massachusetts Institute of Technology in Cambridge.

No written treatment of any subject can ever be complete, of course: application will bring the readers experience. With experience, each reader will decide for themselves what parts of this text are useful and what parts are not, and what has been overlooked or omitted. The purpose of this chapter is to fill some of the gaps that the author pondered during the journey from feedback student to engineering practitioner.

4.1 "But in Control Theory, They Use Lots of Linear Algebra ... "

If this section title sounds funny to you, then you have probably not spent much time among engineering doctoral students. No matter. The point is that in some circles, it would be considered remarkable to get to this point in a "guide" to feedback theory and not once have touched the tools of linear algebra. Matrices, eigenvalues and eigenvectors, singular value decomposition – these form the language of "serious" control theorists.

The reason we have not needed the mathematical machinery of linear algebra is that all of our discussion has concerned systems with *one* input and *one* output. But to start to see the possible need for a new formalism, it helps to

return to Section 1.1.2, where we discussed higher-order systems and remarked
that we can often represent them with systems of linear equations:

$$\frac{dx_1}{dt} = a_{11}x_1 + \cdots + a_{1n}x_n, \tag{4.1}$$

$$\frac{dx_2}{dt} = a_{21}x_1 + \cdots + a_{2n}x_n,$$

$$\vdots$$

$$\frac{dx_n}{dt} = a_{n1}x_1 + \cdots + a_{nn}x_n.$$

Remember that the x_i were the "state" variables in the sense that when know
the value of all of the x_is, we have completely determined the state of the
system.

What we have now in Eq. 4.1 is an undriven dynamical system, or a
system with no input. It starts with initial values for all of the x_is, and the
state dynamically evolves according to Eq. 4.1. But what happens now if we
introduce a single input to the system? We expect that this input will influence
the time evolution of some or all of the state variables. If we label the input
$u_1(t)$, we express this input's influence on the time evolution of the state
variables as

$$\frac{dx_1}{dt} = a_{11}x_1 + \cdots + a_{1n}x_n + b_{11}u_1, \tag{4.2}$$

$$\frac{dx_2}{dt} = a_{21}x_1 + \cdots + a_{2n}x_n + b_{21}u_1,$$

$$\vdots$$

$$\frac{dx_n}{dt} = a_{n1}x_1 + \cdots + a_{nn}x_n + b_{n1}u_1.$$

You actually already know how to handle this case of a single input. If you
were to write Eq. 4.2 as a single, high-order differential equation, you would
wind up with something of the form of Eq. 1.47.

But now, what if the system has *two or more* inputs? It happens. The
dynamics of a fighter jet, or a car for that matter, are shaped by multiple inputs
from the pilot and the environment.[1]

Before diving into the math, it is worth pondering for a moment what you
might do if confronted with analyzing this multiple-input system armed only

---

[1] Control joystick, rudder pedals, and throttle or thruster controls in the case of a jet; accelerator
pedal, brakes, steering wheel in the case of a car; forces due to airflow on and around the body
in *both*.

with the background covered in this book. Interestingly, if you were told that this is a Multiple-Input, Linear Dynamical System that is Stable, you might paralyze yourself trying to figure out how to use linear algebra to derive pole locations. However, if removed from this analytical context, you would almost certainly do an extremely sensible thing: get a "feel" for how the car accelerates without touching the brakes and keeping the steering wheel fixed, then see how the car steers when going a constant speed, then investigate the braking distance with your foot off the accelerator and the steering wheel held constant, and so on. Mathematically, this is investigating the system one input at a time with all the other inputs either zeroed out or constant. While it is possible to miss some dynamic behavior taking this "simplified" approach, the experienced practitioner learns to *never* remove such "simple" methods from his or her toolkit.

Back to math. How might the system of equations change to reflect the presence of multiple inputs? If we have two inputs, it looks like this:

$$\frac{dx_1}{dt} = a_{11}x_1 + \cdots + a_{1n}x_n + b_{11}u_1 + b_{12}u_2, \qquad (4.3)$$

$$\frac{dx_2}{dt} = a_{21}x_1 + \cdots + a_{2n}x_n + b_{21}u_1 + b_{22}u_2,$$

$$\vdots$$

$$\frac{dx_n}{dt} = a_{n1}x_1 + \cdots + a_{nn}x_n + b_{n1}u_1 + b_{n2}u_2.$$

It is at this point that we start looking for a formalism to spare us from so much writing. This is not all laziness, although there is some of that. But if the purpose of hand analysis is to aid in understanding, it will help to make the writing as simple and clear as possible to aid in focusing on the concepts.[2] To this end, the formalism of linear algebra comes into play. We start by defining a vector of state variables

$$\vec{x} = \begin{bmatrix} x_1 \\ x_2 \\ \vdots \\ x_n \end{bmatrix} \qquad (4.4)$$

---

[2] Keep your eye on this. The novice will sometimes approach a compact formulation and complain that the details are "hidden." This is a valid, understandable complaint. The right sequence is to first wrestle with the complicated, ungainly formulation until the details are fully understood. With time, the compact formulation, or shorthand, will start to be a relief.

and two matrices. The first matrix, A, is the connection between the state variables x_i and their time derivatives. It contains the a_{jk}s of Eq. 4.3:

$$A = \begin{bmatrix} a_{11} & a_{12} & \cdots & a_{1n} \\ a_{21} & a_{22} & \cdots & a_{2n} \\ \vdots & \vdots & \vdots & \vdots \\ a_{n1} & a_{n2} & \cdots & a_{nn} \end{bmatrix}. \tag{4.5}$$

The second matrix contains the B coefficients In the case of two inputs, it would look like

$$B = \begin{bmatrix} b_{11} & b_{12} \\ b_{21} & b_{22} \\ \vdots & \vdots \\ b_{n1} & b_{n2} \end{bmatrix}. \tag{4.6}$$

Ultimately this leaves us rewriting Eq. 4.3 as the compact

$$\frac{d}{dt}\vec{x} = A\vec{x} + B\vec{u}. \tag{4.7}$$

Nice, right? We typically go one step further and explicitly define a vector of *outputs*, $\vec{y}$, in terms of the system state and inputs

$$\vec{y} = C\vec{x} + D\vec{u}, \tag{4.8}$$

where C and D are suitably dimensioned matrices.

And this is where your control theory friends take off. They are concerned about the General Case, in which a system can have a huge number of inputs, outputs, and state variables. Are the available inputs sufficient to fully control the system? Are the available outputs sufficient to make a complete observation of the system's internal state? These are the questions that they ask, and the properties of A in particular (such as its eigenvalues and eigenvectors) assume central importance.

It is a worthy subject for those of you who are interested in complex systems. Be assured that if you decide to study this subject, the insight that you have gained from the careful examination of single-input, single-output systems will serve you well and greatly aid your understanding.

4.2 The Problem of "Sinusoids Running Around Loops"

A major focus of this book has been to guide readers to unify the intuitive concepts of feedback theory with the mathematical analysis. However, we treat in this section a particular unification that almost every student of this subject gets disastrously wrong.

Disastrously wrong.

This particular misunderstanding grabs hold of some people so strongly that, by mid-career, they simply cannot release themselves. Let's get into it.

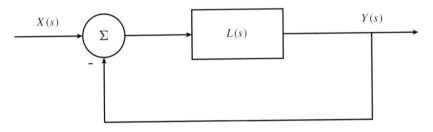

Figure 4.1 A generic feedback loop for discussing the oscillation condition.

At issue is the intuition behind the oscillation condition. That is, for the system in Figure 4.1, we know that it will oscillate at any frequency ω_0 for which

$$|L(j\omega_0)| = 1 \tag{4.9}$$
$$\angle L(j\omega_0) = (2n + 1) \cdot 180°.$$

So what is the error? In trying to explain this condition, an extremely common explanation goes as follows:

Oh, I get it. The sine wave starts out going through $L(s)$ where it gets flipped 180°. Then it gets flipped *again* by the minus sign at the summing junction, so now it is in phase and it just reinforces itself. The sinusoid runs around and around in the loop indefinitely.

If this quote makes no sense to you, stop now, congratulate yourself, and move on immediately to Section 4.3. Don't give it any more thought.

For the rest of us, there are two problems with this explanation. The first problem is going to sound pedantic but is actually quite deep: this conceptual picture mixes time-domain statements with frequency-domain statements. When we invoke a system function like $L(s)$ and work in the frequency domain, we are in the domain of frequencies, and concepts of time as a succession of events do not enter into the picture. That is, when we say that the Fourier transform of a signal is

$$\frac{1}{j\omega + j\omega_0}, \tag{4.10}$$

we mean that its representative back in the time domain is

$$e^{-j\omega_0 t}. \tag{4.11}$$

To be very clear, this is a sinusoid of infinite extent in time that "started" *before* the Big Bang and will still be there to greet the Four Horsemen of the Apocalypse.

When you have that infinite extent in your mind, you have to realize that there is no sensical way to talk about such a signal as "running around a loop." Running around a loop, chasing one's tail – these are thought pictures that only work in a discretized, time-sequenced conceptual framework that has a beginning and an end.

The second problem is that such thinking generalizes poorly. The error is given an unfortunate assist by the language of phase margin: if you have zero phase margin, you get an oscillation; therefore, you might think, if you go even further and have gain greater than unity when the phase is $-180°$, you have a worse problem and the envelope of the oscillation will grow exponentially.[3]

A counterexample can be a powerful corrective. Behold:[4]

$$L(s) = \frac{k(\tau s + 1)^2}{s^3}. \tag{4.12}$$

We start by looking at a Bode plot in Figure 4.2, and calculating the phase margin. According to the strict definition of phase margin, a feedback loop with this loop transmission ($k = 10^5$ and $\tau = 0.1$) would be stable, would have its closed-loop poles safely in the left-half plane. And – as it does in almost all practical cases – the strict phase margin check gives the correct result! But how can this be when, as the Bode plot shows, there is a gain of approximately 100 when the phase is at $-180°$? Can this be right?

We have doubts, and so we fall back on the full Nyquist criterion and make sure. First, we must deal with the awkwardness of having poles at the origin and therefore on the D contour, since these points on the contour would now map to an infinity. Common practice is illustrated in Figure 4.3. We then revert to the convention that the normalized loop transmission, $L_0(s)$, is $\frac{(\tau s+1)^2}{s^3}$.

---

[3] Note that this thinking stretches beyond the strict definition of phase margin given in Section 3.7.2.

[4] Having multiple poles at the origin is not as outrageous or contrived as it may seem at first. One pole at the origin of the loop transmission guarantees zero steady-state error in response to a step input; two poles at the origin does the same for a ramp input; and three gives zero steady-state error for a quadratically rising or falling input. (You can use the final value theorem to prove these results.) Phase-locked loops are one extremely common class of electronic systems where ramp tracking ability is an important consideration.

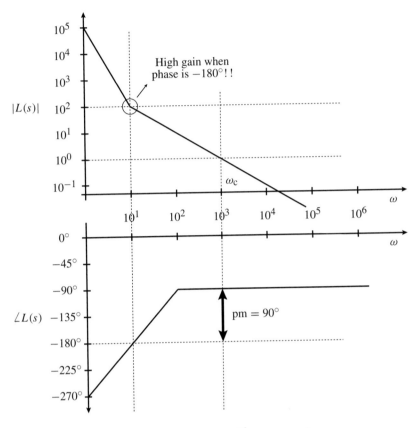

Figure 4.2 A Bode plot for $L(s) = \frac{k(\tau s + 1)^2}{s^3}$ with $k = 10^5$ and $\tau = 0.1$.

We can still make use of the Bode plot of Figure 4.2 if we recall that we must slide the magnitude plot down by a factor of 10^5 since we want to use the normalized loop transmission. We see quickly in Figure 4.4 that the Nyquist criterion confirms this perhaps surprising result. With $k = 10^5$, we are in fact in a region where there are no net encirclements of the $-1/k$ point.[5]

Finally, the root locus for this famous loop transmission is shown in Figure 4.5. This further confirms that for sufficiently high values of k, all poles are safely in the LHP and we have a stable system.

---

[5] For any who still think that Nyquist plots are inconvenient, notice how we got *quantitative* stability results with easy hand analysis. The only other recourse was to find the roots of a third-order polynomial. Not as difficult as it once was in this age of lightning-fast computational aids, but still.

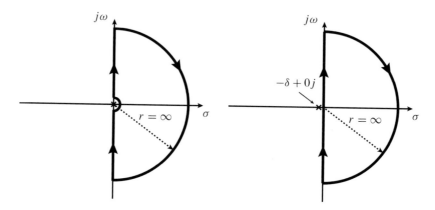

Figure 4.3 Applying the Nyquist criterion when there are poles at the origin. Common practice is shown on the left, where the D contour is modified to have a tiny indentation to keep the contour from actually hitting the poles. On the right is an alternative, where one concedes that the infinite DC gain of a true pole at the origin is unrealizable, and so we shift those poles very slightly into the LHP.

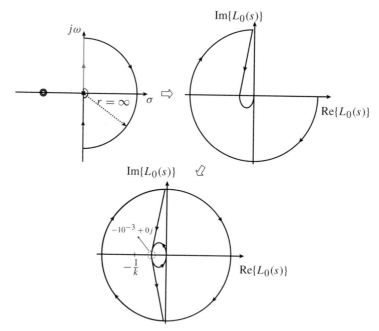

Figure 4.4 Applying the Nyquist stability test to $L(s) = k\frac{(\tau s+1)^2}{s^3}$. As an intermediate step, we plot first half of the Nyquist plot using the positive $j\omega$ axis as shown. Note that the Bode plot allows us to pick out a key numerical feature of the Nyquist plot, which is that the -180 point has an approximate magnitude of 10^{-3}. We can use this to see if our k value is large enough to ensure no net encirclements of the $-1/k$ point.

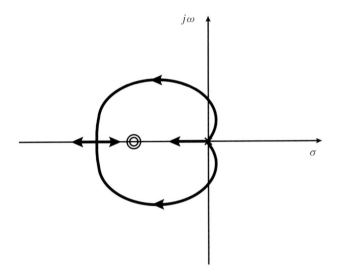

Figure 4.5 Root locus plot: $L(s) = \frac{k(\tau s + 1)^2}{s^3}$.

There are two key results here. First, despite the phase margin relying on a very small fraction of the information contained in the frequency response of the loop transmission, for almost all practical cases it *does* correctly and usefully indicate stability for feedback systems when applied strictly according to its definition. Second, we now have a concrete example wherewith to dismantle this picture of "sinusoids running around in a loop," since in this case we had a gain of 100 where the phase was $-180°$.

Fine. What *is* a useful picture to carry around in one's head?

One useful conceptualization for the oscillation condition comes from the idea of resonance, which crops up often in physical systems. Examples include: a taut string suspended between two supports; or a metal cavity containing electromagnetic fields; or a resonant air cavity. In these examples, only certain vibration "modes" are allowed. These modes are "allowed" because wave patterns at certain frequencies are able to satisfy inviolable "boundary conditions." For a concrete example, consider the taut string anchored at both ends depicted in Figure 4.6. If one plucks the string, the string will vibrate as a superposition of traveling, sinusoidal waves. What these waves all have in common is that an integer number of half-wavelengths fit, exactly, between the fixed supports, making them consistent with the *boundary condition* that the string is fixed in position at $x = 0$ and $x = L$.

In feedback systems it is quite natural, and *proper*, to think of the oscillation condition as a resonance phenomenon. We can imagine the relevant boundary

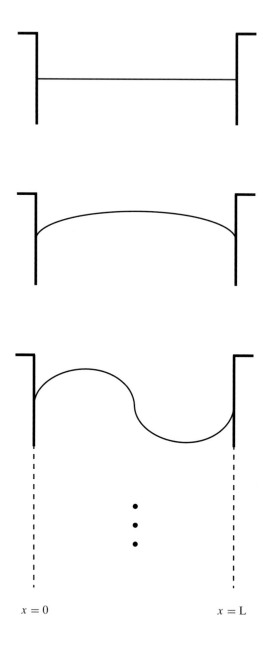

$x = 0$ $x = L$

Figure 4.6 A taut string suspended between two supports. The only oscillatory modes allowed are ones for which an integer number of half-wavelengths fits between $x = 0$ and $x = L$.

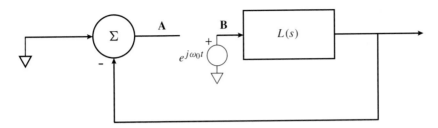

Figure 4.7 Oscillation in feedback systems as satisfying a boundary condition. For a sinusoid to satisfy the oscillatory boundary condition, the signal at point **A** must equal that at point **B**.

condition if we think of breaking the loop as shown in Figure 4.7. In order for an oscillation to occur at frequency ω_0, the *boundary condition* is that in the open-loop configuration of Figure 4.7, the signal at point **A** must be equal to that at point **B**. For a sinusoid, verify that this corresponds exactly to the oscillation condition. Fix in your mind that oscillations are a type of resonance.

4.3 Discrete-Time Control of Continuous-Time Systems

Discrete-time control of continuous-time systems is an incredibly important topic. It is very common for students to immerse themselves in "classical" control theory, often in CT, only to be completely at sea when it comes to using a digital computer for feedback control. The analytical tools that they have developed seem to fit awkwardly, if at all. The purpose of this section is to plug this gap. A DT feedback control system for a CT plant is shown in Figure 4.8.

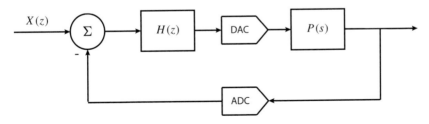

Figure 4.8 DT control of a CT plant. $H(z)$ is the DT compensator, $P(s)$ is the plant, an "ADC" is an analog-to-digital converter, and a "DAC" is a digital-to-analog converter. ADCs sample their analog input and produce digital representations of the input voltage at each sample period. DACs reverse this process, sampling a digital code at their input and producing an analog voltage or current at each sample instant.

DT control of CT systems can also be a very big topic. So how do we constrain things so that a satisfying treatment can be contained in this small volume? A straightforward way to do this is by appealing to a basic fact of analog-to-digital converters (ADCs): they introduce delays.[6] And we now know that delays are deadly in real-time feedback control. So it will turn out that while we *could* do a long, general analysis of DT control of CT systems, what we shall do instead is insist *ahead of time* that delays associated with data converters have a minimal impact on the control dynamics. The result is a dramatic simplification that allows us to employ the full suite of CT analysis techniques with minimal adaptation.

4.3.1 DT Processing of CT Signals

The first step is to determine how to analytically jump from CT to DT and back again. That is to say, we look at Figure 4.9 (bottom) and ask how to express $Y(j\omega)$ in terms of $X(j\omega)$ and $H(e^{j\Omega})$. We can break this down into four substeps:

1. Express $X_d(e^{j\Omega})$ in terms of $X(j\omega)$.
2. Express $Y_d(e^{j\Omega})$ in terms of $X_d(e^{j\Omega})$.
3. Express $Y(j\omega)$ in terms of $Y_d(e^{j\Omega})$.
4. Tie it all together and express $Y(j\omega)$ in terms of $X(j\omega)$.

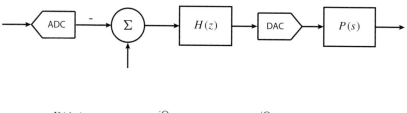

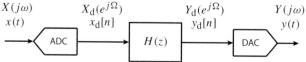

Figure 4.9 (top) "Unrolling" Figure 4.8 to get a clear look at the loop transmission signal chain. (bottom) Same signal chain, boiled down to the essentials for analyzing DT processing of CT signals.

---

[6] Digital-to-analog converters (DACs) also introduce delays.

Details now follow. If you are less interested in how to work this out, the final result is given in Eq. 4.27. In fact, even if you are interested in the details, it might be worth taking a peek ahead at Eq. 4.27. If you look at it carefully, you may decide that this result is exactly what you would expect.

Great. We start with step 1, which is the ADC. It samples $x(t)$ so that $x_d[n] = x(nT)$.[7] And right away, we have an ambiguity to ponder that gets more alarming the more you think about it. The most straightforward way to see the issue is to observe that the sinusoid $x(t) = e^{j\omega t}$ will produce *exactly the same samples* as the sinusoid $x(t) = e^{j\left(\omega + k\frac{2\pi}{T}\right)t}$, where k is an integer. Put another way, a pure sinusoid at frequency f will produce exactly the same samples as a pure sinusoid at frequency $f + kf_s$, where f_s is the sampling frequency $\frac{1}{T}$, and k is a positive or negative integer. It appears that if the original signal is too broadband, it will get mangled by the sampling process.

It is worth a pause in your reading here to internalize this. The essential ambiguity associated with sampling a CT signal is called *aliasing*. For example, imagine a CT input that happens to be periodic with period T_{sig}, and so we can represent it as the Fourier series the way we did in Eq. 1.76:

$$x(t) = A_0 + A_1 \cos\left(\frac{2\pi}{T_{\text{sig}}}t + \phi_1\right) + A_2 \cos\left(2 \cdot \frac{2\pi}{T_{\text{sig}}}t + \phi_2\right) \cdots \quad (4.13)$$

Convince yourself that if the sampling period happened to equal the repetition period of this waveform, $T_{\text{sig}} = T$, then every sinusoid term of Eq. 4.13 would "alias" to DC in the frequency domain and $x(nT)$ would be a constant value![8]

With a sampler, then, we have it that we cannot tell the difference between a frequency component at f and a frequency component at $f + kf_s$. *The fix is to agree ahead of time not to introduce signals at the input of the sampler that fall outside the range* $-\frac{f_s}{2} < f < \frac{f_s}{2}$. Sometimes this condition is enforced by an "anti-aliasing filter," which is a low-pass filter whose cutoff is at or below f_s. It is essential for DT processing of CT signals that we *bandlimit* the CT signals in this way. When we sample too slowly to prevent aliasing, we say that we are "undersampling." Conversely, if we sample faster, sometimes much faster, than what is required to avoid aliasing, we say that we are "oversampling."

---

[7] The ADC also *quantizes* $x(t)$. We will assume here that the quantization level (number of bits) is enough to have no effect on the behavior of the system. For those of you who are concerned: if you know the *effective number of bits* (ENOB) of the converter, you may carry around in your head that the maximum SNDR of the digitized signal is SNDR $= 6.02 \times$ ENOB $+ 1.76$.

[8] Hint: substitute $t = nT_{\text{sig}}$ into Eq. 4.13.

Having taken care of the aliasing issue, we can mostly proceed in peace to get $X_d(e^{j\Omega})$ in terms of $X(j\omega)$. With $\omega_s = 2\pi f_s$, we write the original samples in terms of the CT Fourier transform $X(j\omega)$:

$$x_d[n] = x(nT) = \frac{1}{2\pi} \int_{-\omega_s/2}^{\omega_s/2} X(j\omega)e^{j\omega T n} d\omega. \qquad (4.14)$$

We are a mere change of variables[9] away from the familiar DT Fourier transform of Eq. 1.151. Let $\omega T = \Omega$, which means $\frac{1}{T}d\Omega = d\omega$. For the limits of integration, we have $\frac{2\pi f_s \cdot T}{2} = \pi$, and write

$$x_d[n] = \frac{1}{2\pi} \frac{1}{T} \int_{-\pi}^{\pi} X\left(j\frac{\Omega}{T}\right) e^{j\Omega n} d\Omega. \qquad (4.15)$$

Finally, we complete the first step of the process by writing

$$X_d(e^{j\Omega}) = \frac{1}{T} X\left(j\frac{\Omega}{T}\right). \qquad (4.16)$$

Step 1, finished!

Step 2 is easy:

$$Y_d(e^{j\Omega}) = H(e^{j\Omega})X_d(e^{j\Omega}) \qquad (4.17)$$

$$Y_d(e^{j\Omega}) = \frac{1}{T} H(e^{j\Omega})X\left(j\frac{\Omega}{T}\right).$$

Step 3 is writing $Y(j\omega)$ in terms of $Y_d(e^{j\Omega})$, or capturing mathematically the operation of a DAC. Most DACs operate exactly the way you would think: for each sample $y[n]$, it just holds the correct output voltage (or current) until the next sample. The result is that the abstract sequence of lollipops that we usually use to depict DT signals becomes a staircase voltage waveform Figure 4.10.

Fine. Mathematically, we write the output waveform $y(t)$ as

$$y(t) = \sum_n y[n]p(t - nT), \qquad (4.18)$$

where $p(t)$ is a pulse shape known as a "zero-order hold." Recalling the definition of the unit step function $u(t)$ from Eq. 1.123, we write

---

[9] For many of you, it has probably been a while since you saw, or executed, a "mere" change of variables for an integral. The *reason* for changing variables here is to allow us to work in either the DT or CT formalisms with familiar notation (e.g., frequency as ω or f in CT, Ω in DT). To convince yourself that everything is okay, carefully compare Eqs. 4.14 and 4.15 and see that the two definite integrals will always give the same result.

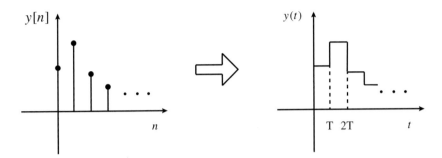

Figure 4.10 Internal to the DT processor, we represent the signal as the sequence of samples on the left. A typical output of a DAC is the staircase waveform on the right, where the output levels are either analog currents or analog voltages.

$$p(t) = u(t) - u(t - \mathrm{T}).$$
(4.19)

We're driving toward getting $Y(j\omega)$ now, so we take the Fourier transform of Eq. 4.18. Remember from Eq. 2.24 that the Fourier transform of $p(t - n\mathrm{T})$ is $P(j\omega)e^{-j\omega n\mathrm{T}}$. $Y(j\omega)$ is therefore

$$Y(j\omega) = \sum_n y[n]P(j\omega)e^{-j\omega n\mathrm{T}}$$
(4.20)

$$= P(j\omega) \sum_n y[n]e^{-j\omega n\mathrm{T}}.$$

Notice that the part of $Y(j\omega)$ left under the summation, $y[n]e^{-j\omega n\mathrm{T}}$, is actually periodic in ω. That is, the summation evaluated at ω_0 is the same if we evaluate it at $\omega_0 + k\frac{2\pi}{T}$, which is $\omega_0 + k\omega_s$, where k is a positive or negative integer.[10] Otherwise, notice that the part under the summation is just $Y_d(j\Omega)$ with $\Omega = \omega\mathrm{T}$.

For $P(j\omega)$, we have

$$P(j\omega) = \int_{-\infty}^{\infty} [u(t) - u(t - \mathrm{T})]e^{-j\omega t}\,dt$$
(4.21)

$$= \frac{1 - e^{-j\omega\mathrm{T}}}{j\omega}.$$

---

[10] Lest you think this to be mere theoretical trivia: if you look at the output of a real-life DAC on a spectrum analyzer, you *will* see spectral "replicas" spaced out at multiples of the sampling frequency. This is a fundamental artifact of sampling, and we battle it in the real world with a combination of filtering and the highest practical sampling rate.

Now we're going to do a few more steps that will look like an arbitrary math dance but will actually give additional clarity:

$$\frac{1 - e^{-j\omega T}}{j\omega} = e^{-j\omega T/2}\left(\frac{e^{j\omega T/2} - e^{-j\omega T/2}}{j\omega}\right) \tag{4.22}$$

$$= e^{-j\omega T/2}\left(\frac{2\sin \omega T/2}{\omega}\right)$$

$$= e^{-j\omega T/2}\left(\frac{\sin \omega T/2}{\omega/2}\right)$$

$$P(j\omega) = Te^{-j\omega T/2}\left(\frac{\sin \omega T/2}{\omega T/2}\right).$$

The multiplicative term $e^{-j\omega T/2}$ shows that a DAC with a zero-order hold imposes a half-sample delay in the outgoing signal stream. This makes sense if you think about it: what you would really want from a zero-order hold is for the *middle* of the pulse to coincide with sample instant nT, and therefore for the pulse to start a time $T/2$ earlier than it actually does. Also, notice that we have placed a factor of T out front to cancel the factor of T placed in the denominator. This little nicety makes the expression in parentheses the classic *sinc* function, in this case sinc$(\omega T/2)$.

We are near the end. We have for $Y(j\omega)$

$$Y(j\omega) = Te^{-j\omega T/2}\text{sinc}(\omega T/2)Y_d(e^{-j\omega T}). \tag{4.23}$$

Substituting now for $Y_d(e^{-j\omega T})$, we have

$$Y(j\omega) = Te^{-j\omega T/2}\left(\frac{\sin \omega T/2}{\omega T/2}\right) \cdot \frac{1}{T} \cdot X_{\text{periodized}}(j\omega)H(e^{j\omega T}). \tag{4.24}$$

Two bits of housekeeping. We will, of course, cancel the factors of T. Next, $X_{\text{periodized}}(j\omega)$ is just $X(j\omega)$ with replicas placed at integer multiples of the sampling frequency as required by Eq. 4.20:

$$X_{\text{periodized}}(j\omega) = \sum_k X(j(\omega + k\omega_s)). \tag{4.25}$$

However, in a practical system, we will assume that we are not aliasing when we sample the inputs and that DAC output replicas are attenuated through a combination of high sample rate and low-pass filtering. If we take care of these details, we arrive at the coveted result of passing a CT signal through a DT filter:

$$Y(j\omega) = X(j\omega)H(e^{j\omega T}) \cdot \text{sinc}(\omega T/2) \cdot e^{-j\omega T/2}. \tag{4.26}$$

Make one final observation, which is that fast sampling compared to the frequency components of $X(j\omega)$ corresponds to $\omega T = 2\pi f/f_s \ll 1$, which means Eq. 4.26 becomes

$$Y(j\omega) \approx X(j\omega)H(e^{j\omega T}). \qquad (4.27)$$

It pretty much had to work out this way. In words, what Eq. 4.27 says is, "Map the frequency components of X onto the $j\Omega$ axis according to $\Omega = \omega T$; apply the filtering function H as you normally would; then return to CT." The rest of Eq. 4.26, the delay term and the sinc term, assert themselves when the sampling frequency is no longer large compared to the highest-frequency components in X. For real-time control, we shall see that you are better off oversampling. And the more the better.

4.3.2 Don't Kid Around: Just Oversample

Having worked out how to jump from CT to DT and back again, we turn our attention to the system at hand. Again, Figure 4.8 is a representative block diagram. Concerned as we are with the dynamics of feedback control, we immediately move to the relevant loop transmission, shown in Figure 4.9 (top). Out of respect for the headaches that even small delays can cause, we are going to rely on the more exact expression in Eq. 4.26 and write the loop transmission as

$$L(j\omega) = H(e^{j\omega T}) \cdot \text{sinc}(\omega T/2) \cdot e^{-j\omega T/2}. \qquad (4.28)$$

It is helpful now to introduce a modification to the loop transmission based on the behavior or real-world data converters. Even the fastest ADCs introduce delay between the input and output: some of the fastest, so-called pipelined ADCs explicitly exploit "latency," or delay, in exchange for a high sample rate. Other data converters, such as successive-approximation ADCs, introduce latency in exchange for a higher degree of quantization. For many applications, such as digital communications, a delay of even a few hundred samples simply does not matter. But for real-time control, these delays are of critical importance. We will therefore modify our loop transmission according to

$$L(j\omega) = H(e^{j\omega T}) \cdot \text{sinc}(\omega T/2) \cdot e^{-j\omega T\left(\frac{1}{2}+n_{\text{ADC}}+n_{\text{DAC}}\right)}. \qquad (4.29)$$

We come now to the critical point, which is, what do we consider "fast" sampling? The author suggests that "fast" sampling is sampling fast enough that the phase impact of the delay terms in Eq. 4.29 are no greater than that

of a *a pole that is a decade above the highest frequency of interest*.[11] It turns
out that a single, real pole exerts a phase pull of $-5.6°$ at a frequency that is
one-tenth of its magnitude. So we require

$$\omega T\left(\frac{1}{2}+n_{ADC}+n_{DAC}\right)\le\left(\frac{5.6°}{360°}\right)\cdot 2\pi\approx 0.0977\text{ rad},\qquad(4.30)$$

which means, for all frequencies ω of interest,

$$\omega T\le\frac{0.0977\text{ rad}}{\frac{1}{2}+n_{ADC}+n_{DAC}}.\qquad(4.31)$$

A numerical example will save this whole development from being too
abstract. At the time of this writing, Analog Devices Inc. offered a low-latency
ADC with an advertised latency of 2.5 samples. Low-latency DACs are more
common, again at the time of writing, but even assuming no extra latency for
the DAC we still have the half-sample delay from Eq. 4.26. Assuming we are
using these super low-latency data converters, then, we have that

$$\omega T\le\frac{0.0977\text{ rad}}{3}\approx 0.0326.\qquad(4.32)$$

We can go back and check that we have satisfied our assumptions by looking
at the real and imaginary parts of the delay term at this outer limit where
$\omega T=0.0326$:[12]

$$e^{-j\omega T(\frac{1}{2}+n_{ADC}+n_{DAC})}=e^{-j(0.0326)\cdot\left(\frac{1}{2}+2.5\right)}\qquad(4.33)$$

$$=e^{-j(0.0977)}$$

$$=\cos(0.0977)-j\sin(0.0977)$$

$$=0.995-j0.0975.$$

If you check, the angle of the delay term is indeed $-5.6°$. Even better, the sinc
term works out to be satisfyingly close to 1:

$$\text{sinc}(\omega T/2)=\frac{\sin(0.0326/2)}{0.0326/2}=9.999557\times 10^{-1}\approx 1.\qquad(4.34)$$

We conclude that if we sample fast enough to keep data converter artifacts from
introducing significant delays and phase shifts, we justify the approximation
of Eq. 4.27. In a practical control system, it is reasonable to assert that the
highest frequency "of interest" is the unity crossover frequency of the loop

[11] A natural "frequency of interest" in control systems is the unity crossover frequency, where we
measure phase margin.
[12] If you're checking these calculations with your own calculator, remember that the arguments
of sin and cos are in radians.

transmission, $\omega_c = 2\pi \cdot f_c$. Applying our criterion for keeping sampling delays from disrupting the loop dynamics, we find that we require the sampling frequency, $f_s = 1/\text{T}$, to be

$$\omega_c \cdot \frac{1}{f_s} \leq 0.0326 \tag{4.35}$$

$$f_s \geq \frac{2\pi f_c}{0.0326}$$

$$f_s \geq 193 \cdot f_c.$$

That is to say, we are oversampling by quite a lot! Those of you familiar with signal processing might recall the theoretical minimum sample rate to avoid aliasing, which would have been

$$f_s \geq 2 \cdot f_c. \tag{4.36}$$

But remember that delay is *bad* for real-time feedback control. Regardless of how elaborate your analysis gets, there is simply no causal tool for fighting delay when we sit down to do feedback compensation. The best policy is to architect your system from the start to prevent large phase shifts from delays. In the case of DT control of CT systems, this means heavily oversampling.

4.3.3 Relationship between z and s in Mixed-Signal Control

We have already established that, for DT processing of CT signals, and when we do enough oversampling, the loop transmission is well approximated by $L(j\omega) \approx H(e^{j\omega \text{T}})$ (Eq. 4.27). When we consider now the design of DT compensators, it is helpful to check that this "rule" generalizes to

$$L(s) = H(e^{s\text{T}}). \tag{4.37}$$

Can we do this?[13] To find out, we lower our head and turn the crank, dusting off the rarely used full form of the inverse Laplace transform, which we last saw in Eq. 1.113:

$$x(n\text{T}) = \frac{1}{2\pi j} \int_{\sigma_0 - j\infty}^{\sigma_0 + j\infty} X(s) e^{s\text{T}n} ds. \tag{4.38}$$

We mentioned before that this is a contour integral. The contour along which we integrate is the vertical line in the complex plane for which the real part

---

[13] If your first reaction is, "Of course, why wouldn't we be able to?" well, you are correct. Does not hurt to check, though. You can perhaps concede that through the first few lines of the derivation, anyway, this final result is conceivably in doubt.

of s is σ_0. The only restriction on σ_0 is that it must be a value for which the original Laplace transform converges.

It turns out that that way through the wilderness here is by doing three changes of variables. Ultimately, we want to make Eq. 4.38 turn into something like an inverse Z-transform of the samples $x[n]$. That is, we are looking to end with an expression like

$$x[n] \propto \oint X(\cdot)z^{n-1}dz. \tag{4.39}$$

Now if you compare Eqs. 4.38 and 4.39, you may despair of getting from one to the other. But recall that z is a complex variable and that when we are in DT, we tend to write it as $z = re^{j\Omega}$. Try blindly substituting this form of z into Eq. 4.39, and recall that we just finished a development where we found the equivalence $\Omega = \omega T$. There are many details to fill in, which we will do in the following, but do that simple substitution to see that Eqs. 4.38 and 4.39 maybe aren't as far apart as they look at first.

The first change of variables for Eq. 4.38 will be $s = \sigma_0 + j\omega$. For the contour of integration, σ_0 is a constant, and so for purposes of the change of variables $ds = j \cdot d\omega$. We are now integrating $X(\sigma_0 + j\omega)$ and sweeping the variable ω, so in order to keep the result of the integration the same the limits on the definite integral must change to $-\infty$ and $+\infty$. The steps look like the following:

$$x(nT) = \frac{1}{2\pi j} \int_{\sigma_0 - j\infty}^{\sigma_0 + j\infty} X(s)e^{sTn}ds \tag{4.40}$$

$$= \frac{1}{2\pi j} \int_{-\infty}^{+\infty} X(\sigma_0 + j\omega)e^{(\sigma_0 + j\omega)Tn} j \cdot d\omega$$

$$= \frac{1}{2\pi} \int_{-\infty}^{+\infty} X(\sigma_0 + j\omega)e^{(\sigma_0 + j\omega)Tn}d\omega.$$

Finally, if we assume that there is no aliasing then $X(\sigma_0 + j\omega)$ is zero for ω outside of the range of $[-\omega_s/2, \omega_s/2]$, and so we finish this first change of variables by writing

$$x(nT) = \frac{1}{2\pi} \int_{-\omega_s/2}^{+\omega_s/2} X(\sigma_0 + j\omega)e^{(\sigma_0 + j\omega)Tn}d\omega. \tag{4.41}$$

The second change of variables is $\omega T = \Omega$. We aim to sweep Ω as the variable of integration. To keep the result of the definite integral unchanged we will: replace $d\omega$ with $\frac{1}{T} \cdot d\Omega$; the argument of $X(\cdot)$ inside the integral becomes $\sigma_0 + j\frac{\Omega}{T}$; and, crucially, the limits of integration become $\pm\frac{\omega_s}{2} \cdot T$.

But $\omega_s = 2\pi f_s = \frac{2\pi}{T}$. So the actual limits of integration become $\pm\pi$, and we have

$$x(nT) = \frac{1}{2\pi}\frac{1}{T}\int_{-\pi}^{+\pi} X\left(\sigma_0 + j\frac{\Omega}{T}\right)e^{\sigma_0 Tn}e^{j\Omega n}d\Omega. \qquad (4.42)$$

We will do the third and final change of variables after making a substitution $r = e^{\sigma_0 T}$:

$$x(nT) = \frac{1}{2\pi}\frac{1}{T}\int_{-\pi}^{+\pi} X\left(\sigma_0 + j\frac{\Omega}{T}\right)(re^{j\Omega})^n d\Omega. \qquad (4.43)$$

Now the final change of variables: $z = re^{j\Omega}$, and it will be important to recall that r is fixed as we do our integral because σ_0 is fixed:

$$dz = jre^{j\Omega}d\Omega, \qquad (4.44)$$

$$dz = jzd\Omega,$$

$$d\Omega = \frac{1}{j}z^{-1}dz.$$

The final bit of trickery is, as always, to keep the result of the integral the same as part of doing the change of variables $z = re^{j\Omega}$. We can do this by changing this contour integral to a *new* contour integral, this time with a circle in the z-plane with radius $r = e^{\sigma_0 T}$. Since we go from $-\pi$ to π it is a complete circle. Substituting in for $d\Omega$, and of course replacing $re^{j\Omega}$ with z, we wind up with

$$x[n] = x(nT) = \frac{1}{2\pi j}\frac{1}{T}\oint X\left(\sigma_0 + j\frac{\Omega}{T}\right)z^{n-1}dz. \qquad (4.45)$$

This is none other than the Z-transform that we sought from the beginning. The final result is that

$$X_d(z) = \frac{1}{T}X\left(\sigma_0 + j\frac{\Omega}{T}\right), \qquad (4.46)$$

together with the important relations

$$\sigma_0 = \frac{1}{T}\ln|z| \qquad (4.47)$$

$$\Omega = \arg z.$$

At long last, we have the Z- and Laplace transform analogs to the Fourier transform relation of Eq. 4.16. Notice that if we stick to the unit circle in the z domain, $\sigma_0 = 0$ and Eqs. 4.16 and 4.46 are equivalent, exactly as they should be. In fact, what we really have here is a generalized mapping from s to z and back again. When we have taken care to avoid aliasing, we can in fact treat the s and z variables on either side of the data converters through the relations

$$z = e^{sT} = e^{\sigma T}e^{j\omega T} \tag{4.48}$$

and the reverse

$$\sigma = \frac{1}{T}\ln|z| \tag{4.49}$$

$$\omega T = \arg z = \Omega.$$

We go through all of this to generalize Section 4.3.1, and we generalize because we want to describe how to analytically design DT compensators for CT systems in Section 4.3.4. Fortunately, steps 2–4 for the generalization work out way more easily than step 1 did.

Step 2:

$$Y_d(z) = H(z)X_d(z), \tag{4.50}$$

where $X_d(z)$ is defined in Eq. 4.46.

Step 3, writing $Y(j\omega)$ in terms of $Y_d(z)$:

$$y(t) = \sum_n y[n]p(t - nT) \tag{4.51}$$

$$Y(s) = P(s)\sum_n e^{-sT}$$

$$= P(s)[Y_d(z)]|_{z=e^{sT}}$$

$$= P(s)H(e^{sT})X_d(e^{sT}).$$

We drive toward the generalization of Eq. 4.26 with

$$Y(s) = \frac{1}{T}P(s)H(e^{sT})X(s). \tag{4.52}$$

$P(s)$ is the zero-order hold $\frac{1-e^{-sT}}{s}$, so

$$Y(s) = \frac{1}{T} \cdot \frac{1 - e^{-sT}}{s} \cdot H(e^{sT})X(s). \tag{4.53}$$

We complete this analysis by looking in the oversampling limit, for which $|sT| \ll 1$ and thus $e^{-sT} \approx 1 - sT$. In the oversampling limit, we arrive at

$$Y(s) \approx H(e^{sT})X(s), \tag{4.54}$$

as we hoped we would. This result allows for a relatively simple analytical framework for designing feedback compensators in mixed signal systems, as we shall see in the next subsection.

4.3.4 DT Compensators for CT Systems

Where does this all lead for the design of DT compensators for CT systems? A very simple approach is to do your compensator design as if you were staying in CT. After you have chosen a compensator $H(s)$, take an inventory of your poles p_i and zeros z_i. For every desired pole in $H(s)$ you place a pole at $e^{p_i T}$ in the DT compensator, and similarly place your zeros at $e^{z_i T}$.

Note that the major consequence of strong oversampling is that we are in the limit of DT as described in Section 2.5.5. That is, all of our poles and zeros are concentrated near the $1 + 0j$ point on the unit circle. An interesting weirdness of being in DT, though, is the possibility of placing unit sample delays in the transfer function. As we have emphasized, there are so many reasons to avoid this. But it will happen that most realizable transfer functions in DT will come with one or more sample delays. Recall that the Z-transform of a unit delay is z^{-1}; these delays will show up as poles at the origin of the z-plane. These special poles do not translate one-to-one to equivalent poles in CT. Rather, we must tally them up and tack on a multiplicative e^{-sTm} term to our CT equivalent, where m is an integer equal to the total number of DT poles at the origin.

4.3.5 The Other Useful Extreme: Slow Sampling

Up to this point in Section 4.3, we have argued strongly the merits of oversampling. That is, the sample interval T is extremely short compared to the dynamics of the plant that is being controlled.

But it is not always true that high speed of response is the most important design goal. In some control applications, it is sufficient for the machine to slowly step to its final value, and all that matters is that it get there eventually, and with high accuracy. In such applications, the DT controller may make a step in the right direction, poll all relevant sensors to see how close it is to the final goal, and then take another step. This is a situation where the sample interval T is extremely long compared to the dynamics of the plant. During a sample interval, the plant has time to completely settle, so that from the standpoint of the DT controller it appears to be infinitely fast. In this situation, one may regard the problem as a pure DT feedback control problem and need not worry about crossing back and forth between CT and DT analysis domains.

4.3.6 A Note on the Bias toward CT Methods

The readers will note that CT analysis methods are given a privileged place in the preceding analysis. The primary reason for this is that CT analysis tools tend to be a little friendlier to pencil-and-paper analysis, which strengthens understanding and intuition before optimizing with computer tools.

But different people have different preferences, and the best designs come when the individual designer is convinced that their method is the *best*. So if you are drawn more to DT methods and design tools, the good news is that the $z = e^{sT}$ mapping described in Section 4.3.4 still works for you. You will need to model the CT element that is being controlled. Then for each of its poles and zeros, you will map them to their DT equivalents in your control model. From there, analyze and design the way that works best for you.

4.3.7 Sometimes, Real-Time Computer Control Is Hopeless

Many times, despite targeting high sampling rates for the data converters, the sample delays in DT systems add up and become overwhelming. The practice of "pipelining" digital signal paths is near-universal, and normally for good reason: the impact of propagation delays in complex logic paths would greatly reduce achievable clock rates unless latency was actively traded for speed. And then there is the realization of digital filters themselves ($H(z)$). Figure 4.11 shows a canonical DT integrator, the simplest of DT filters. A look at the transfer function

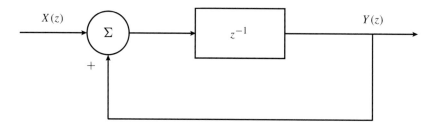

Figure 4.11 A typical DT integrator. Notice that neither input to the summing junction is inverting. The governing difference equation is $y[n+1] = x[n] + y[n]$.

$$H(z) = \frac{z^{-1}}{1 - z^{-1}} = z^{-1} \cdot \frac{1}{1 - z^{-1}} \tag{4.55}$$

reveals that in addition to the desired pole at the $1 + 0j$ point, we also get a unit sample delay z^{-1} that is most undesired. Although there are forms of the DT integrator that lack this particular drawback, it is often true that digital filters are beset by a substantial number of sample delays.

It would seem that there is no way out, that the situation is hopeless. That is ... true. But if the plant we are controlling has characteristics that only

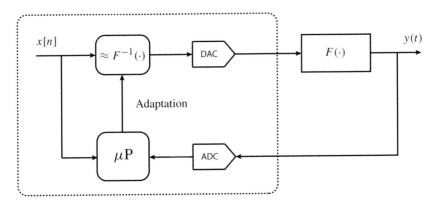

Figure 4.12 Adaptive predistortion. In these systems, "stability" relates to the ability of the system to stably converge to an accurate inverse model $F^{-1}(\cdot)$. A digital microprocessor (μP) manages the adaptation.

vary slowly, we can do astonishingly well by building a system like that shown in Figure 4.12. In this strategy, the digital controller uses feedback data to adapt an "inverse model" of the plant. This is essentially open-loop control that learns from past mistakes, and it is immensely powerful. It is how virtually every cellular base station in the world today implements "adaptive, digital predistortion." The delays in the signal path have an impact only on how quickly we are able to converge on an inverse model, and more delay will in fact increase convergence time. But because the transmit chain in base stations change characteristics due to factors like temperature changes and aging, we can slow down adaption as far as we need to accommodate convergence without ever sacrificing needed throughput.

5

Feedback in the Real World

A nice thing about feedback theory in textbooks is its relative neatness and simplicity. In a block diagram, for example, it does not matter how many inputs the output of a given block is driving. In the real world, "loading effects" are where the behavior of a system block may be altered depending on what its output is connected to. Also true in abstract treatments is that the only connections present are the ones that you see. There is no such thing as an unmodeled coupling between one block and another.[1]

All this means is that when you go out to apply feedback theory to the real world, you cannot expect to accomplish its application simply by rote. It will require flexibility and creativity and will benefit from deepening experience, just like any other rewarding, high-level discipline. Learn to embrace this, if you have not already. Rote application would be boring!

5.1 Finding Loop Transmissions

Consider a machine that we will call a "cooling unit." The idea is that in response to an input voltage, the cooling unit acts to cool a room at a rate proportional to an input voltage. Suppose further that the unit is capable of "negative cooling." That is, it heats the room if the input voltage is negative.

Now, we said that the rate of cooling/heating is proportional to the input voltage. This means that the temperature, T_R, is related to the input voltage of the cooling unit v_{in} according to

$$\frac{dT_R}{dt} \propto -kv_{in}, \tag{5.1}$$

---

[1] Or between the output of one block and the input of that same block, which is an unintended feedback loop.

or equivalently,

$$T_R \propto -k \int v_{in} dt. \qquad (5.2)$$

Here k is a positive constant that, in this simplified model, captures such effects as the size of the room (larger rooms have smaller ks) and the coupling of the cooling/heating element to the air. Details for now. The point here is that we are now ready to draw a block diagram of a feedback system for controlling the temperature of the room. We see that the forward path is going to have a pole at the origin, so as a first pass, we'll try nothing more complicated than proportional control. Our controller, then, is the simple gain block of Figure 5.1, and our "plant" is the combination of the cooling unit and the thermal dynamics of the room, shown illustrated in Figure 5.2. Let us not forget that we are reading the room temperature with an electronic sensor that converts temperature to voltage via a proportionality constant k_T as shown in Figure 5.3.

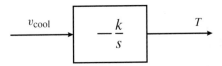

Figure 5.1 Proportional control for the temperature system.

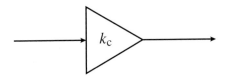

Figure 5.2 Capturing the relationship between input voltage to the cooling/heading unit and room temperature.

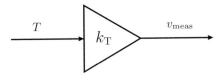

Figure 5.3 The temperature sensor converts temperature to voltage with some proportionality constant k_T.

We are ready to connect it all together on paper and consider the question of stability. Going from rote memory of feedback theory, we might draw something like Figure 5.4.

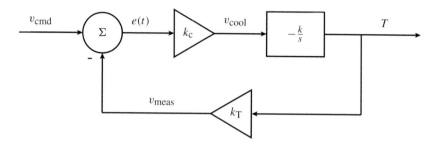

Figure 5.4 A simple control loop for the temperature of a room.

5.1.1 Is the Sign Right? A Useful Check

But before diving into all of the details of analysis, it is useful to do a simple check to see if the sign of the loop transmission is right. In the author's experience of designing and building feedback systems over many years, this simple check has proven time and again to be a most useful way to help get one's mind around the inner workings of the system. The check is, is the sign of the loop transmission right? That is, in response to a discrepancy between the desired temperature and the measured temperature, does the system act to *reduce* instead of *increase* the discrepancy?

The way to check this is to look at the open-loop system resulting from "breaking apart" the original closed-loop system. As shown in Figure 5.5, this open-loop system has two inputs, v_{cmd} and v_{meas}. And now, we do a very simple check. Suppose that the temperature in the room is lower than the desired temperature, so that we have $v_{cmd} - v_{meas} > 0$. In this case, $e(t)$ will be a positive voltage, as will v_{cool}. You can see that the integrator will attempt to drive its output, which is actually the temperature in the room, *lower and lower as long as the condition $v_{cmd} - v_{meas} > 0$ persists*. This is exactly what we do not want. It's the behavior of a right-half-plane pole!

The sign of the loop transmission is wrong. Without doing root locus or Nyquist or any other stability analysis, a simple check reveals that what we first wrote down is backward and makes no sense. Fortunately there is an easy fix: make $k_c < 0$.

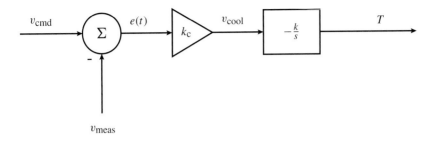

Figure 5.5 Checking the sign of the loop transmission.

Now, it is true that this mistake would ultimately be caught by any of the stability analysis techniques we have discussed in this book. It is also true that getting the sign of the loop transmission right is no guarantee of stability. But getting the sign wrong *does* almost guarantee instability. Keep an eye on this in your own designs. It is a great way to stay grounded in the physical behavior of the system.

5.2 A Common Application: Howling Speakers and Microphones

The readers will not find it difficult to find applications of feedback theory in the real world. Applications abound in the author's speciality of electronic circuits: op-amp circuits, phase-locked loops, $\Sigma\Delta$ modulators, and oscillators (both intentional and accidental) are but a start to a very, very long list. It is impossible to do anything but scratch the surface of potential applications in a single book, much less a single chapter.

So scratch the surface we shall, with an example that is familiar to almost everyone: the howling public address system. These systems consist of a microphone, an amplifier, and a loudspeaker and are intended to allow a single person to address a large crowd without having to shout. You know the scenario: at just about every wedding reception, a member of the wedding party stands up to make their remarks and kicks things off by tapping the microphone and asking, "Can you all hear me?" The system faithfully responds with an ear-splitting howl, and the guests duck and cover their ears. The moment is saved when someone shouts, "Turn down the volume!" Which is just another way of saying, "Lower the gain in the loop transmission of that feedback loop!" Let's examine.

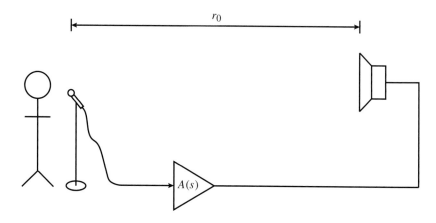

Figure 5.6 Microphone, amplifier, and loudspeaker of a public address system.

The beginnings of a plausible model are sketched in Figure 5.6. We can see the potential issue right away: the output of the speaker depends on both the intended input (the person's voice) *and* the output of the speaker itself. It certainly is a feedback system. How do we determine its stability?

To answer this, we must map out a reasonable model of the loop transmission.[2] We can start with the audio amplifier itself. Since the purpose of the audio amplifier is to amplify only signals that are in the audio range with reasonable flatness, we might suppose that it has a transfer function with a single pole at 20 kHz. The transfer function is sketched in Figure 5.7.

Thinking about the assumptions that we have made, we may well feel satisfied that we have a reasonable framework for an amplifier. But we find that we do not immediately have a number to plug in for the gain of the amplifier, and the assumption that it has a single pole is reasonable but is still a fairly big assumption. Must we give up so soon after starting?

This is a great example of needing to use creativity and judgment in a real-world application. One approach at this point could have been to stop the wedding reception, dismantle the amplifier, get data sheets for all of the components, and really get a detailed model on paper. Another approach is

---

[2] Sometimes students are hesitant to commit to a mathematical form for a model. What if, they suppose, they get it wrong? Rest assured: you are *practically guaranteed to get it wrong*! This is true for everyone! The point is to commit, honestly and carefully, your full understanding to a model, and then check the model's validity through experiment. If your model turns out to be correct, great! If not, the process is to understand the behavior that your model failed to capture, modify your model accordingly, *and* thoughtfully analyze why you made the errors that you did. Repeat as many times as necessary There may be no other way to build real understanding.

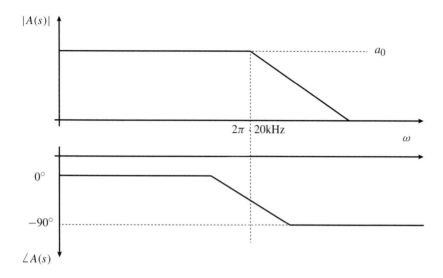

Figure 5.7 A Bode plot for the amplifier in Figure 5.6.

to proceed with what we have, noting that our model already has analytical features that may give us insight if we pursue things further. While we shall go with the latter approach for now, this is a legitimate judgement call that will be either validated or repudiated as we take next steps. Keep a sharp eye out for the ego's involvement here: it is easy, and paralyzing, to get too hung up on being "right" on all of these little decisions along the way. Feel your way through, and learn from your mistakes rather than fear them.

We press ahead, inserting variables as placeholders for values we don't know in the transfer function of the amplifier $A(s)$:

$$A(s) = \frac{a_0}{\tau s + 1}, \tag{5.3}$$

where τ is $\frac{1}{2\pi \cdot 20\text{kHz}}$. At this point we might look up and wonder a little. After all, if the sign of the loop transmission works out in our favor, there is no way to get oscillations with only a single pole in the transfer function.[3] It must be true that there are physically important effects that are we have not modeled.

It turns out that we must include two effects when we model the coupling between the speaker output and the microphone input. The first effect could be that the speed of sound through air is about 343 m/s, so there is a delay term in the loop transmission. But should it matter? We can do a quick check. Suppose

---

[3] Verify this for yourself with your analytical tool of choice: algebra, root locus, Nyquist, phase margin.

the microphone and the speaker are separated by a relatively short 3 meters (approx. 10 feet). This equates to a delay of 9 ms. A sound wave right in the heart of the audio range has a frequency of 1 kHz, or a period of 1 ms. So even this modest separation between the speaker and the microphone introduces a whopping $9 \times 360° = 3240°$ of phase! We definitely expect this delay to be important.

The second effect that our experience says makes a difference is that separating the microphone and the speaker seems to help things. While this certainly adds even more phase from the delay that we just discussed, it is also true that the coupling between microphone and speaker is weakened. Sound dies away, after all, the farther you are from the source, and happens to do so in an inverse-square manner. To within a constant, we may model this as a gain block that depends on r_0, the separation between microphone and speaker:

$$A_A = \frac{k_A}{r_0^2}. \tag{5.4}$$

If we put all of this together, we have a model of the feedback loop shown in Figure 5.8. The summing junction is the microphone, which combines the intended input (the voice of the person speaking) with the unintended input (the output of the loudspeaker).

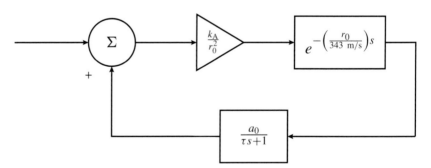

Figure 5.8 A block diagram model for the public address system.

We'll do one more thing, which is has to do with the fact that the summing junction here has positive signs associated with both inputs, in contrast to the usual way things are done. How you deal with this is up to you. The author's personal preference is to make it a minus sign, and insert an $e^{-j\pi}$ block in the feedback path, as shown in Figure 5.9.

Fantastic – the hard part is actually done. Once you have the problem captured as neatly contained blocks in a diagram, the stability analysis goes quickly, and all the more so if you've had a lot of practice. Now we just have

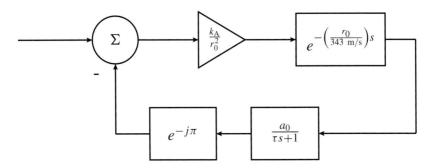

Figure 5.9 Same system as in Figure 5.8, but force fitting it into the negative feedback framework that we have used throughout this book.

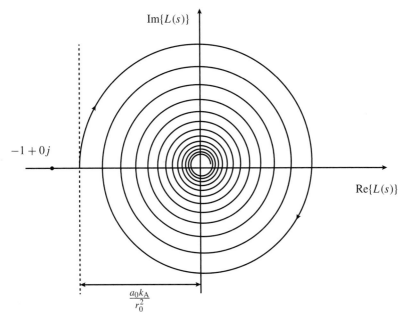

Figure 5.10 A parameterized half Nyquist plot (positive frequencies) for Figure 5.9. The 90° phase shift from the amplifier is in there somewhere. But for audio frequencies, this phase will get utterly swamped out by a propagation delay for r_0 greater than a few centimeters.

to maintain our flexibility, since we can't draw *all* the details of the Nyquist plot without exact values for k_A, r_0, s_0, and a_0.[4] A parameterized Nyquist plot is shown in Figure 5.10.

---

[4] This is actually better in the early stages of understanding the physics of a system. Later, when we want to nail down a design, the numerical values of the various parameters become more important.

What do we see? Our main concern is whether we will see oscillations, that horrible howling that these systems are famous for, when we turn on the power. What we see from the Nyquist plot is that if a_0, the amplifier gain, is too big, and/or the distance between the microphone and speaker is too small, the Nyquist contour will indeed expand beyond the $-1+0j$ point, and we will have what we dread. So our model, as rough as it is, manages to capture analytically what we know from experience: when the howling starts, we should either turn down the volume (lower a_0) or move the speaker and the microphone farther apart (increase r_0).

6

Conclusion and Further Reading

Well, there you have it: *A Guide to Feedback Theory*, and along the way a walk back through the linear systems that are the mathematical foundation of your entire engineering study. Think deeply about these topics, and in so doing, do not forget to play and have fun. Follow your curiosity, and try to work out answers to your own questions as they come up. When you get stuck, talk about your stuckness with like-minded people, search the internet or your engineering library for answers, or, surprisingly, walk away for some time, and let your subconscious chew on the problem for a while.

What follows is a list of books that had a great impact on me as I learned and pondered feedback topics over the last 35 years. I was an undergraduate in the EECS department at MIT in the mid- to late 1990s, which this list certainly reflects. Otherwise, what you will find here are not just clear expositions of the stated subject materials, although they are certainly that. These books always struck me as extremely personal, where the authors shared their peculiar takes on widely discussed subjects.

Feynman, Richard, Robert Leighton, and Matthew Sands. *The Feynman Lectures on Physics*. Reading, MA: Addison-Wesley, 1963.

Halliday, David, and Robert Resnick. *Fundamentals of Physics*. 3rd ed. Hoboken, NJ: John Wiley, 1988.

Hildebrand, Francis B. *Advanced Calculus for Applications*. Upper Saddle River, NJ: Prentice Hall, 1976.

Kailath, Thomas. *Linear Systems*. Upper Saddle River, NJ: Prentice Hall, 1980.

Karu, Zoher Z. *Signals and Systems Made Ridiculously Simple*. Huntsville, AL: ZiZi Press, 1995.

Lee, Thomas. *The Design of CMOS Radio-Frequency Integrated Circuits*. Cambridge, UK: Cambridge University Press, 1993.

Roberge, James K. *Operational Amplifiers: Theory and Practice.* New York, NY: John Wiley, 1975.

Siebert, William M. *Circuits, Signals, and Systems.* Cambridge, MA: MIT Press, 1986.

Strang, Gilbert. *Introduction to Linear Algebra.* Wellesley, MA: Wellesley-Cambridge Press, 1993.

It is worth emphasizing one more time: remember to play, whether it be with pencil and paper, your favorite simulator, MATLAB, or whatever. If a derivation in this book intrigues you or seems important, pay attention to that curiosity. Close the book, take out a blank sheet of paper, and see how much of the derivation you can recover on your own. When you get stuck, just remember that getting stuck usually exposes a lack of understanding as opposed to an inability to remember. Think deeply about where you are stuck and why, address the misunderstanding, and then try again on a new sheet of paper. This is a great way to deepen your understanding.

Index